THE
NEUTRINO

BY

JAMES S. ALLEN

PRINCETON UNIVERSITY PRESS
PRINCETON, NEW JERSEY
1958

Copyright, 1958, by Princeton University Press
L. C. Card 57-5464

Printed in the United States of America

PREFACE

When this monograph was in the early stages of preparation, the neutrino appeared to be a respectable and well-established member of the large family of particles of nuclear physics. The existence of the neutrino was postulated by Pauli in 1933, and this concept was immediately used by Fermi in his theory of beta-decay. The Fermi theory was so successful in the explanation of most of the important features of beta-decay that most physicists accepted the neutrino as one of the "particles" of modern physics. In 1955 the general subject of neutrino experiments and neutrino theory appeared to be closed.

However, the appearance in 1956 of the important paper by Lee and Yang on the question of parity conservation in weak interactions initiated a large number of theoretical investigations and experimental studies of processes which involve the neutrino or anti-neutrino. Although these investigations have revealed new facts concerning the neutrino, the subject now is far from being closed, and it is likely that active research in this field will be continued for several years. In view of these new developments, this monograph should be considered as a progress report of the investigations of the neutrino through May 1957. Although the experimental method has been emphasized, an attempt has been made to interpret the experimental results in terms of existing theories of the neutrino.

The author wishes to acknowledge the continued help and instruction of his colleague, Professor J. Weneser, in the formulation of beta-decay theory. The author is also obliged to Professor D. R. Hamilton of Princeton University for his careful and critical reading of the entire manuscript.

James S. Allen

Urbana, Illinois
May, 1957

TABLE OF CONTENTS

THE NEUTRINO

CHAPTER 1

General Properties of the Neutrino

1.1. Introduction

The neutrino has remained one of the most interesting particles of nuclear physics ever since its existence was postulated by Pauli[1] in 1933. The concept of the neutrino has been so useful in the theory developed by Fermi[2] to explain the beta-decay process that most physicists now accept it as one of the "particles" of modern physics. With the exception of the recent free neutrino experiments of Cowan and his co-workers,[3] the characteristics of the neutrino have been deduced almost entirely from somewhat indirect experimental evidence.

The neutrino when introduced by Pauli was assigned the role of carrying away the missing energy, momentum, and spin in the beta-decay process. Since this new particle had never been observed directly, the assumption was made that it was neutral and had a very small mass. The existence of a continuous spectrum of electron energies from a typical radioactive isotope such as P^{32} was explained by the assumption that the total energy available from the nuclear transformation was shared by the electron, the neutrino, and the recoiling final nucleus. The conservation of linear momentum was also assured by the assumption that the momentum of the recoiling nucleus was equal and opposite to the vector sum of the momentum of the electron and that of the neutrino.

In many examples of beta-decay, angular momentum apparently is not conserved. Consider the beta-decay, $C^{14} \rightarrow N^{14} + e^- + \nu$. The nuclear angular momenta of C^{14} and N^{14} have been measured; they are 0 and 1, respectively. Thus, the nuclear angular momentum changes during the transition by an integral amount $\Delta J = 1$. The electron with an intrinsic spin of $\frac{1}{2}$ cannot conserve the angular momentum. However, the addition of the neutrino with an intrinsic spin of $\frac{1}{2}$ will solve this difficulty. A spin of $\frac{3}{2}$ would also allow momentum to be conserved. However, the calculations of Kusaka[4] appear to exclude this choice on the basis of the shape of the beta-spectrum.

1.2. Neutrinos and Anti-Neutrinos

When describing the neutrino it is necessary to consider the possibility that there may be an anti-neutrino which is distinguishable from the neutrino. The anti-neutrino may be conceived as a "hole" in the

3

sense used for positrons. We may assume that the neutrino and anti-neutrino can be described by a Dirac[5] equation of the type used for the electron and positron. However, if the neutrino is described by a modified equation due to Majorana,[6] no anti-particle exists.

All beta-decay processes may be expressed in terms of the fundamental relations:

$$n \rightarrow p + e^- + \nu^* \tag{1.1}$$

and

$$p \rightarrow n + e^+ + \nu. \tag{1.2}$$

Equations (1.1) and (1.2) can be considered as the definitions of the anti-neutrino, ν^* and the neutrino, ν. The introduction of the anti-neutrino is convenient in the formulation of the theory of beta-decay. For example, the probability of a transition resulting in the emission of a negative electron and an anti-neutrino can be considered to be proportional to the square of matrix elements of the form:

$$\int (\psi_p^\dagger O \psi_n)(\phi_e^\dagger O \phi_\nu) \, d\tau, \tag{1.3}$$

where the wave functions (field amplitudes) $\psi_p^\dagger$, $\phi_e^\dagger$ correspond to the created particles, and ψ_n, ϕ_ν correspond to the absorbed particles. The light particle wave functions usually are four-component spinor solutions of the Dirac relativistic equation, and the operator O is a 4×4 Dirac matrix (see Bethe[7] or Schiff[8] for the explicit form of the Dirac matrices). In principle the transforming nucleons should also be described by relativistic wave functions. The daggers indicate that the Hermitian conjugate of a given matrix is to be used. With similar reasoning the matrix elements which appear in positron emission are of the form:

$$\int (\psi_n^\dagger O^\dagger \psi_p)(\phi_\nu^\dagger O^\dagger \phi_e) \, d\tau, \tag{1.4}$$

where now the wave functions $\psi_n^\dagger$, $\phi_\nu^\dagger$ correspond to the created particles and ψ_p, ϕ_e correspond to the absorbed particles. Thus, negative electron emission can be associated with the absorption of a neutrino or the emission of an anti-neutrino. Positron emission can be associated with the absorption of a negative electron from the filled negative energy states and the emission of a neutrino.

Until very recently both the neutrino and the anti-neutrino, if indeed these are different particles, appeared to have the same physical properties, namely: a spin of $\frac{1}{2}$, no electrical charge, and an extremely small or perhaps zero magnetic moment and mass. The distinction between these two particles appeared rather artificial due to this apparent identity in their properties. However, the recent suggestion of Lee and Yang[9] that parity may not be conserved in weak interactions such as the beta-decay interaction has led to new theoretical investigations and experimental proof that the two particles are distinguishable.

If parity is not conserved then the angular distribution of the electrons coming from the beta-decays of oriented nuclei should be asymmetrical with respect to the angle between the orientation of the parent nucleus and the momentum of the electron. Fortunately there is experimental evidence which indicates that there is an asymmetry in the angular distribution of the electrons coming from the beta-decays of oriented nuclei, and consequently parity conservation is violated in beta-decay. The violation of parity conservation implies the existence of a right-left asymmetry. In the case of the beta-decay, the neutrino apparently exhibits this asymmetrical behavior with respect to the right and the left.

Lee and Yang[10] have also proposed a two-component theory of the neutrino which is possible if parity is not conserved. In this theory for a given momentum q the neutrino has only one spin state, the spin being always parallel to q. The spin and momentum of the neutrino together automatically define the sense of a screw. The neutrino may then be defined as a particle with spin parallel to its momentum, and the anti-neutrino as a particle with spin anti-parallel to its momentum. The fact that this description of the neutrino does not conserve parity (special inversion) can be explained as follows: under a spacial inversion P, the momentum of the neutrino is inverted, but not the direction of its spin. Since in this description the two are always parallel, the operator P applied to a neutrino state leads to a nonexisting state. Consequently the theory is not invariant under space inversion.

In this new theory the mass of the neutrino must be zero and its wave function need have only two components instead of the usual four. The need for a zero mass can be explained as follows: unless the mass of the neutrino is identically zero, a moving observer could overtake a neutrino in flight and the momentum of the particle would apparently reverse. Since the direction of the spin would not reverse, this again would lead to a nonexisting state. If the rest mass were identically zero, the neutrino would move with the velocity of light and therefore could not be overtaken by a moving observer.

It is an interesting fact that a two-component theory is not new. This type of theory was suggested by Weyl[11] in 1929 and was examined later by Pauli.[12] However, the theory was always rejected because of its intrinsic violation of space inversion invariance—a reason which is no longer valid.

A more detailed discussion of these new theories of the neutrino will be presented in Chapter 4. Experimental evidence concerning the conservation of parity will be discussed in Chapter 5. It should be emphasized that at present the available experimental evidence does not definitely prove that the neutrino must be described entirely in terms of the two-component theory. A description based upon a mixture of both parity-conserving and nonconserving interactions may prove to be correct.

In general the use of either the Dirac or the Majorana descriptions of the neutrino leads to the same end results when applied to single beta-decay. However, we shall see in Chapter 2 that the exact shape of the high-energy end of the electron spectrum does depend on the description of the neutrino. Unfortunately this effect vanishes if the rest mass of the neutrino or anti-neutrino is zero. Since the results of recent experiments indicate that the mass of the neutrino is essentially zero, practically no information regarding the description of the neutrino can be obtained from the study of single beta-decay.

The subject of double beta-decay will be discussed in Chapter 6. In this type of decay two neutrons are transformed into two protons with the emission of two electrons. If two Dirac anti-neutrinos must be emitted in this transition, a half-life of the order of 10^{20} years is expected for a kinetic energy release of 2 Mev . The transformation is greatly speeded up if a Majorana neutrino is emitted by one neutron and captured by the other so that no neutrinos are actually emitted. The expected half-life would then be of the order of 10^{12} years. Despite the extreme rarity of double beta-decay, recent experiments in this field appear to indicate that the Dirac picture of the distinguishable anti-neutrino is the correct one. In the following chapters the term "neutrino" will be used to refer to either the neutrino or anti-neutrino unless it is necessary to distinguish between the two particles.

1.3. Methods for the Indirect Observation of the Neutrino

There are a number of reactions which involve the neutrino, and experiments based on some of these have yielded rather detailed information regarding the properties of this particle. In these experiments the neutrino has not been observed directly, but indirect evidence of its presence has been obtained from the momentum transferred to a recoiling nucleus,

An upper limit for the rest mass of the neutrino has been deduced from experimental investigations of the shape of the continuous beta-spectrum near the end point. In the interpretation of this type of experiment the assumption is made that the Fermi theory uniquely explains the shape of the beta-spectrum and that, as a result, the shape near the upper energy limit is sensitive to the mass of the neutrino. A more detailed description of these experiments will be given in Chapter 2.

Experimental investigations of the disintegration of nuclei by electron capture clearly indicate that one neutrino is emitted in this process. According to our present picture of the orbital electron capture process, an electron from one of the atomic shells is captured by a nucleus according to the scheme:

$$(Z + 1)^A + e^-_{K,L} \rightarrow Z^A + \nu, \tag{1.5}$$

where Z^A represents a nucleus of atomic number Z and mass number A. If the binding energy of the orbital electron is neglected compared to the nuclear energies, and if atomic masses are used, the energy of the emitted neutrino (or neutrinos) is given approximately by:

$$W_\nu = M(Z + 1) - M(Z). \tag{1.6}$$

The energy spectrum of the recoiling nuclei should consist of a single line if the single neutrino picture is correct. In contrast, the emission of multiple neutrinos should result in a continuous distribution of recoil energies. A detailed discussion of several important orbital electron capture experiments will be presented in Chapter 3.

Originally many beta-decay processes were successfully investigated in order to show that conservation of energy and linear momentum requires the neutrino hypothesis. In recent years the emphasis has been on electron-neutrino angular correlation experiments. These experiments are based on the fact that the shape of the energy or of the momentum spectrum of the recoiling nuclei is a relatively sensitive function of the interaction or interactions responsible for beta-decay. A detailed description of the theory, the experimental procedures, and the interpretation of the results of angular correlation investigations will be presented in Chapters 4 and 5.

Two important meson reactions should be included in any discussion of indirect observations of the neutrino. These are the decay of the π meson given by:

$$\pi^\pm \rightarrow \mu^\pm + \nu, \tag{1.7}$$

and the decay of the μ meson:

$$\mu^\pm \rightarrow e^\pm + 2\nu. \tag{1.8}$$

In principle, the neutrino rest mass may be deduced from reaction (1.7). The decay of the μ meson given by equation (1.8) is closely related to the ordinary beta-decay of nuclei, and the shape of the spectrum should give some information regarding the interactions responsible for this decay process.

1.4. Direct Observation of the Neutrino

The complete identification of a new particle is always aided by an experimental observation of some direct interaction of this particle with known particles. In the case of the neutrino the search for a direct interaction has extended over a period of about twenty-nine years, beginning with the unsuccessful absorption experiment of Ellis and Wooster,[13] and ending with the successful inverse beta-decay experiment of Cowan, Reines, Harrison, Kruse, and McGuire.[3]

Bethe and Bacher[14] have pointed out that the so-called inverse beta-process is the only action of free neutrinos which can be predicted with certainty. According to this inverse process, a neutrino is captured by a nucleus accompanied by the emission of a positive or a negative electron. The cross-section calculated for this interaction by Bethe[15] is approximately 10^{-44} cm^2. The cross-section for inverse beta-decay has been successfully measured by Cowan and co-workers,[3] and this experiment will be described in Chapter 7.

A summary of the direct interaction experiments through 1948 is to be found in a neutrino review by Crane.[16] He concluded that all cross-sections greater than 10^{-36} or possibly 10^{-37} cm^2 had been excluded on the basis of these experiments. Since these experiments are of considerable historical interest in the chronological development of the investigations leading to the final successful measurement of a direct interaction, they will be briefly described in Chapter 7.

REFERENCES

1. Pauli, W. in *Rapp. Septieme Conseil Phys.*, Solvay, Brussels (Gautier-Villars, Paris, 1934) (1933).
2. Fermi, E. *Zeits. f. Phys. 88*, 161 (1934).
3. Cowan, C. L. Jr., Reines, F., Harrison, F. B., Kruse, H. W., and McGuire, A. D. *Science 124*, No. 3212, pp. 103–104 (July 20, 1956).
4. Kusaka, S. *Phys. Rev. 60*, 61 (1941).
5. Dirac, P. A. M. *Proc. Roy. Soc. A117*, 610 (1928) and The Principles of Quantum Mechanics, 2nd ed., Chapter XII. Oxford, New York, 1935.
6. Majorana, E. *Nuovo Cimento 14*, 171 (1937).
7. Bethe, H. A. *Handbuch der Physik*, Vol. 24, 1, Chapter III. Julius Springer, Berlin, 1933.
8. Schiff, L. I. *Quantum Mechanics*. McGraw-Hill, New York, 1949.
9. Lee, T. D., and Yang, C. N. *Phys. Rev. 104*, 254 (1956).
10. Lee, T. D., and Yang, C. N. *Phys. Rev. 105*, 1671 (1957); see also Landau, L. *Nuclear Phys. 3*, 127 (1957); and Salam. *Nuovo Cimento 5*, 299 (1957).

11. Weyl, H. *Zeits. f. Phys. 56*, 330 (1929).
12. Pauli, W. *Handbuch der Physik*, Vol. 24, pp. 226–227. Julius Springer, Berlin, 1933.
13. Ellis, C. D., and Wooster, W. A. *Proc. Roy. Soc. A117*, 109 (1927).
14. Bethe, H. A., and Bacher, R. F. *Rev. Mod. Phys. 8*, 82 (1936).
15. Bethe, H. A. *Elementary Nuclear Theory*, pp. 21. Wiley, New York, 1947.
16. Crane, H. R. *Rev. Mod. Phys. 20*, 278 (1948).

REVIEW ARTICLES

Pontecorvo, B. *Rep. Prog. Phys. 11*, 32 (1946–1947).
Crane, H. R. *Rev. Mod. Phys. 20*, 278 (1948).
Allen, J. S. *Am. J. Phys. 16*, 451 (1948).
Sherwin, C. W. *Nucleonics 2*, 16 (May 1948).
Kofoed-Hansen, O. *Physica XVIII*, No. 12, 1287 (1952).
Kofoed-Hansen, O. *Beta and Gamma Ray Spectroscopy*, Chapter XII edited by K. Siegbahn. Interscience, New York, 1955.

CHAPTER 2

The Rest Mass of the Neutrino

2.1. Introduction

One of the fundamental steps in the identification of a new particle is the determination of the rest mass of this particle. If the results of a number of different experiments yield essentially identical values of the mass, we usually assume that the same particle is involved in each of these experiments. In principle the rest mass of the neutrino can be deduced from any reaction in which this particle appears. However, since the mass is less than one percent of that of the electron, the most suitable reactions are those in which the kinetic energies of the emitted particles are of the order of a few kev or less. The most accurate estimates of the mass have been obtained from beta ray experiments and from orbital electron capture experiments. We shall see that the accuracy of the value of the mass determined from beta ray experiments is limited by the uncertainty as to the proper method for the measurement of the maximum kinetic energy of the beta-spectrum. In this chapter we shall confine our attention to beta ray experiments. Orbital electron capture experiments will be discussed in the following chapter.

2.2. The Energy Balance in Beta-Decay

Bethe[1] and Crane[2] have shown that an estimate of the mass of the neutrino can be made from a closed cycle in which a p, n reaction is followed by a positron emission. For example, Haxby, Shoupp, Stephens, and Wells[3] have measured the energy threshold of the reaction:

$$C^{13} + p \rightarrow N^{13} + n; \tag{2.1}$$

$$N^{13} \rightarrow C^{13} + e^{+} + \nu, \tag{2.2}$$

where ν represents the neutrino. In this cycle the masses of C^{13} and N^{13} cancel out. The energy available for the emission of the neutrino and the positron may be found if the n-H mass difference, the mass of the positron, and the mass of the electron are known in addition to the threshold for reaction (2.1). The available energy for the N^{13} decay, as obtained by this method is 1.20 ± 0.04 Mev. When this is compared with the value of 1.198 ± 0.006 Mev obtained by Lyman[4] for the end point of the N^{13} positron spectrum, the energy equivalent of the mass of the neutrino is found to be very small, if not zero. Hughes and Eggler[5] have deduced a value of the neutrino mass from the reactions:

$$N^{14}(n, p)C^{14} \tag{2.3}$$

$$C^{14} \rightarrow N^{14} + e^- + \nu; \tag{2.4}$$

and

$$He^3(n, p)H^3 \tag{2.5}$$

$$H^3 \rightarrow He^3 + e^- + \nu. \tag{2.6}$$

When the n-H mass difference, the mass of the electron, and the energies available for the beta-decay of C^{14} and H^3 were substituted in these closed cycles, the neutrino mass was found to be 1 ± 25 kev from the first cycle, and 4 ± 25 kev from the second. In general the accuracy of the mass of the neutrino as obtained from reactions which depend on the balance of energy in beta-decay is limited by the uncertainty in the measurement of the total energy release in the beta-decay. If the energy release is obtained from a measurement of the maximum beta energy, as shown by Kofoed-Hansen,[6] the accuracy of this value will be no greater than the accuracy desired for the mass of the neutrino. This point will be discussed in greater detail in the following sections of this chapter.

2.3. Beta-Spectra and the Neutrino Mass

The phenomenological definition of the neutrino and Fermi's use of it in his formulation of the theory of beta-decay make no stipulation about the rest mass of the neutrino other than to assume that it is small. Kofoed-Hansen[6] has shown that the precise shape of an allowed beta-spectrum in the vicinity of the upper energy limit depends on the rest mass of the neutrino. However, he omitted a small, relativistic correction term which depends on the rest mass of the neutrino. According to Pruett,[7] this correction term is:

$$1 \mp \frac{\nu}{W(W_0 - W + \nu)}.$$

The momentum distribution of the beta-decay electrons then becomes:

$$N(p) \, dp = \text{const } F(Z, W)p^2(W_0 - W + \nu)$$

$$\times [(W_0 - W + \nu)^2 - \nu^2]^{1/2} \times \left\{1 \mp \frac{\nu}{W(W_0 - W + \nu)}\right\} dp, \tag{2.7}$$

where W_0 is the maximum electron energy including the rest energy, p is the electron momentum corresponding to the energy W, and ν is the rest mass of the neutrino in units of m. All energies are in units mc^2 and all momenta in units mc. The expression inside the braces in equation (2.7) arises when the factor $\left|\{\phi_e^\dagger, O\phi_\nu\}\right|^2$ is summed over the directions of the spins of the electron and neutrino (or anti-neutrino). $\phi_e^\dagger$ and ϕ_ν are the wave functions of the light particles, and O represents an operator characteristic of the beta-decay interaction. The negative sign is used if a Dirac anti-neutrino accompanies negative electron emission. The positive sign would apply to Fermi's original formulation of the theory in which the Dirac neutrino accompanied negative electron emission. If a Majorana neutrino (one having only positive energy states) were emitted in the beta-decay process, the factor depending on ν inside the braces would not appear. A more detailed discussion of the various factors which appear in equation (2.7) will be presented in Section (4.2).

In order to find the maximum energy W_0, a Fermi plot of the beta-spectrum frequently is used. For points not too near the high-energy end of the spectrum ($W_0 - W + \nu \gg \nu$), the Fermi plot of equation (2.7) is given by:

$$\left[\frac{N(p)}{F(Z,\,W)p^2}\right]^{1/2} = \text{const} \times \left[(W_0 - W) + \nu\left\{1 \mp \frac{1}{2W}\right\}\right]. \qquad (2.8)$$

Very near the end of the beta-spectrum the Fermi plot of equation (2.7) is no longer a straight line, but turns sharply towards the energy axis. If we assume that the anti-neutrino is emitted with the negative electron, the distance between the theoretical true end point and that obtained by extrapolation of equation (2.8) will be $\nu/2$, expressed in the appropriate energy units.

We have just seen that the value of W_0 as obtained by the extrapolation of a Fermi plot is uncertain by approximately ν. Therefore the problem of deducing the neutrino mass from the energy balance in beta-decay, as discussed in Section 2, is indeterminable when the energy release is obtained from a measurement of the maximum beta energy. However, since the shape of the beta-spectrum near the upper energy limit should yield information regarding ν, a number of experimental investigations of this region of the spectrum have been carried out. The effect, if any, should be most pronounced in very low-energy spectra such as H^3 and S^{35}.

2.4. The Beta-Spectrum of H^3 and the Mass of the Neutrino

The beta-spectrum emitted in the "super allowed" transition of H^3 to He^3 has an end point of approximately 18 kev, and therefore is especially well suited for determinations of the neutrino mass. This exceedingly low-energy beta-spectrum has been investigated by proportional counter methods. In these experiments tritium gas was introduced into a high multiplication counter. This method partially eliminates the absorption of the electrons which occurs when the source is a thin layer deposited on a suitable support. However, the accuracy of the energy measurements made with a proportional counter is limited by the finite energy resolution of the counter and also by end effects. The energy resolution is determined principally by the finite number of ion pairs produced by the initial ionizing event. The response of this type of counter to a well-defined beam of ionizing particles is a "peak" in the number of ions which is Gaussian in shape. For the measurements near the end point of the H³ spectrum, the standard deviation in the width of the resolution curve is about four percent or 720 ev. The observed shape of the beta-spectrum must be corrected for this effect.

At the ends of a proportional counter the multiplication decreases due to reduced field strength. Ionizing events occurring in this region will produce pulses of spuriously low amplitude. This effect can be investigated by measuring the response to a beam of mono-energetic alpha particles at various points along the counter. In principle, corrections for this effect may be applied to the observed beta-spectrum.

Curran, Angus, and Cockroft[9] have carried out several investigations of the H³ beta-spectrum using a proportional counter. The energy resolution of the counter as measured with the 17.4 kev K_α x-ray line of molybdenum was five percent for the half-width at half-intensity. The authors modified the theoretical energy distribution curves to allow for the spread corresponding to the resolution of their counter. A comparison between the observed curves and the modified theoretical curves provided a discrimination between various end points and masses. For $\nu = 0$ the experimental points lay between the two curves computed for true end points of 18.0 and 18.6 kev. Alternatively, for a true end point of 18.0 kev the experimental points lay between the curves computed for $\nu = 0$ and $\nu = 1$ kev. Since the theoretical curves probably did not include the correction factor shown at the end of equation (2.7), the quoted upper limit of 1 kev for the neutrino mass should be increased to 2 kev.

Hanna and Pontecorvo[8] have also used the proportional counter technique to investigate the beta-spectrum of H³. The linearity of their counters was checked directly by measuring the Mo-K_α/A³⁷ pulse size ratio. The x-rays resulting from the decay of A³⁷ through orbital electron capture gave a 2.8 kev calibration line. The energy resolution curves of the two counters used in these experiments had standard deviations of about 1 kev at the 17.4 kev calibration point. Fig. (2.1) shows the experimental and corrected points obrained using the better of the two counters. A correction has been made for the finite energy resolution of the counter. The dotted curve has been drawn to show the effect of a neutrino rest mass of 1 kev, with the assumption that an anti-neutrino is emitted in the decay of H³. Apparently neither the experimental points nor the theoretical curve were corrected for the finite resolution

of the counter when the mass was assumed to be 1 kev. The absence
of this correction prevents a critical comparison between the theoretical
and experimental curves, and as a result it is difficult to set an associated
upper limit for the mass of the neutrino.

Langer and Moffat[10] have carried out a detailed investigation of the
H^3 spectrum with a shaped magnetic field spectrometer having a 40-cm

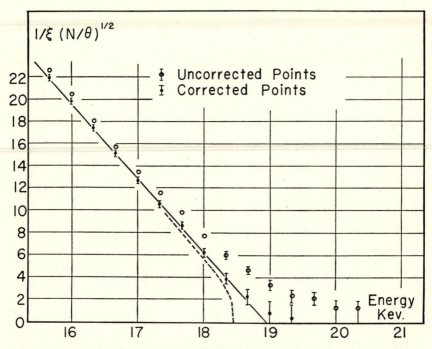

Fig. 2.1. Fermi plot of the end of the H^3 spectrum obtained with a proportional
counter. The corrected experimental points may be compared with the solid line
which corresponds to the distribution expected if the rest mass of the neutrino is zero.
The dotted curve has been drawn to show the effect of a neutrino rest mass of 1 kev
with the assumption that an anti-neutrino is emitted in the decay of H^3. A critical
comparison between the dotted curve and the experimental points is not possible
since neither the experimental points nor the theoretical curves were corrected for
the counter resolution. From Hanna and Pontecorvo.[8]

radius of curvature. This spectrometer was operated with a momentum
resolution of 0.7 percent for the full width at half-maximum. The
corrections applied to points recorded within two percent of the maximum energy amounted to about 0.2 percent in momentum. The calibration of the momentum scale of the spectrometer was made in terms
of the "*A*" line of thorium "*C*" which is known with high accuracy and
lies near the end point of the H^3 spectrum.

The nature of this experiment required that the source be extremely thin and uniform. Tritiated succinic acid with a specific activity of 0.9 mc/μg was chosen as the source material, since this compound is quite stable at room temperature. Sources were prepared by evaporating the succinic acid upon a 4 μg/cm² Zapon support. The evaporated sources were estimated to be about 0.5 μg/cm² thick. Two methods were used for grounding the sources to prevent electrostatic charging. In the first method a thin band of copper was evaporated in vacuum onto the reverse side of the Zapon support. This band extended beyond the edge of the source so as to make contact with the metal ring which supported the source. The resistance of the copper deposit was sufficiently low to insure good grounding. In the second method a source of thermal electrons was placed just below the H³ source in the spectrometer. It was expected that any tendency of the source to charge positively would be neutralized by the low-energy electrons. Each grounding method alone, as well as the combination of copper backing plus low-energy electrons gave spectra with identical end points. With no provision for grounding, the potential of the source increased to a positive value of about 460 volts, as indicated by the shift in the end point of the beta-spectrum.

A Fermi plot of the data in the region near the end point is shown in Fig. (2.2). The small magnitude of the resolution correction required in the immediate neighborhood of the end point is shown by the displacement of the circles to the positions of the crosses. The authors conclude that their data are consistent with a Dirac neutrino (or anti-neutrino) rest mass of zero, and set an upper limit of 250 ev for a possible finite mass, assuming that an anti-neutrino is emitted in the decay process.

Hamilton, Alford, and Gross[11] have investigated the end of the H³ spectrum by means of an electrostatic beta spectrometer. In this spectrometer electrons from a source at the center of a spherical symmetry passed through a retarding field, and those having more than a certain energy reached a collector thus giving an integral spectrum. The current to the collector was sufficiently large to be measured with a direct current amplifier. The energy resolution curve of this instrument had a full width at half maximum of 0.67 percent or about 120 ev at the end point of the H³ spectrum.

The source used in these experiments consisted of tritium absorbed in a layer of 100 μg/cm² of zirconium deposited on a tungsten button. The authors showed that the differential energy spectrum was integrated over the last kilovolt near the end point as a result of the thickness of the source. Since the electrostatic spectrometer is inherently an integrating device, the final result was that the collector current at a given retarding voltage represented a double integral of the beta-spectrum. Assuming that the differential beta-spectrum for zero neutrino

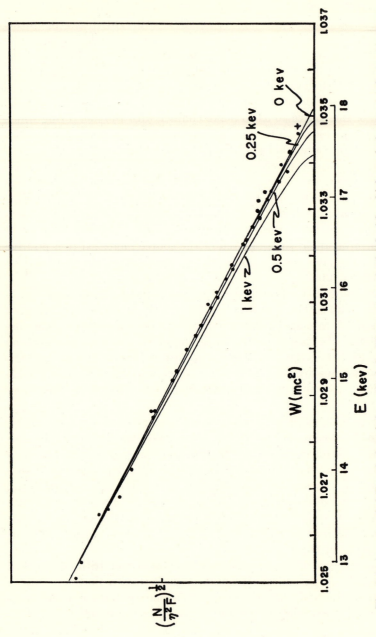

Fig. 2.2. An expanded Fermi plot of the end of the H³ spectrum obtained with a high resolution magnetic spectrometer. The data are consistent with a neutrino rest mass of zero, and set an upper limit of 250 ev for a possible finite mass, assuming that an anti-neutrino is emitted in the decay process. From Langer and Moffat.[10]

mass was proportional to $(W_0 - W)^2$ near the end point, the collector current was expected to be proportional to the fourth power of the retarding voltage.

The double integral spectrum of H³ near the end point is shown in Fig. (2.3). The dotted curves have been computed on the basis of the measured resolution of the spectrometer for various neutrino masses and interactions. It is to be noted that curve D for zero neutrino mass extends beyond the extrapolated end point as a result of the finite energy resolution of the spectrometer. The authors conclude that curves B, C, and D are consistent with their experimental data, and that curve A is inconsistent with the data. If a Dirac anti-neutrino is emitted during the beta-decay, the upper limit for the neutrino mass is about 500 ev.

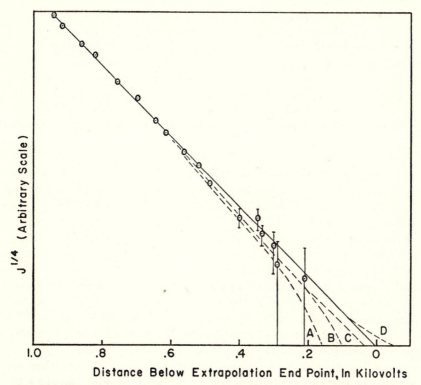

Fig. 2.3. The "double integral" spectrum of H³ near the end point. Dotted curves have been plotted on the basis of the measured resolution and for various neutrino masses and interactions. Majorana, Dirac anti-neutrino, and neutrino interactions are indicated by (0), (−), (+), respectively. The neutrino masses are in electron volts.

Curve A: $\nu = 250$ (+), 350 (0).
Curve B: $\nu = 150$ (+), 200 (0).
Curve C: $\nu = 500$ (−).
Curve D: $\nu = 0$ (+, −, 0).

From Hamilton, Alford, and Gross.[11]

2.5. The Beta-Spectrum of S^{35} and the Mass of the Neutrino

Cook, Langer, and Price[12] have investigated the end of the S^{35} spectrum with the same magnetic spectrometer which was used later for the H^3 experiments. S^{35} decays by negative electron emission, and the end point of the spectrum is at $W_0 = 1.33\ mc^2$, corresponding to a kinetic energy of about 169 kev. In comparison with the H^3 end point of about 18 kev, the higher energy of the S^{35} spectrum is less favorable for neutrino mass determinations. The authors conclude from their experimental results that the rest mass of the neutrino is less than one percent of the mass of the electron, that is, less than 5 kev. As usual, the assumption is that a Dirac anti-neutrino is emitted during the beta-decay.

We may summarize the results of the various investigations of beta-spectra near the end points with the statement that the rest mass of the neutrino certainly is less than 1 kev and probably is less than 250 ev. A more accurate determination of the mass will be extremely difficult, partly because of the need for higher energy resolution, and partly due to the uncertainty in the correct interpretation of the shape of the spectrum near the end point. However, we have progressed one step further in the identification of the neutrino in terms of its fundamental properties or characteristics.

The phenomenological definition of the neutrino as used in Fermi's theory of beta-decay makes no stipulation regarding the rest mass of the neutrino. In the conventional theory the assumption is made that the neutrino can be described by the usual four-component Dirac equation. Lee and Yang[13,14] have recently suggested that space inversion (parity) may not be invariant in weak interactions such as beta-decay. They have also examined a two-component theory of the neutrino which is possible[15] if parity is not conserved. In this theory, for a given momentum p, the neutrino has only one spin state: the spin always parallel to p. The spin and the momentum of the neutrino together define the sense of a screw. According to this description the neutrino could be defined to have a right-handed screw motion, and similarly, the anti-neutrino would have a left-handed screw motion.

According to this two-component theory the mass of either the neutrino or the anti-neutrino is necessarily zero. In addition, since the neutrino and anti-neutrino states cannot be the same, a Majorana theory for this type of neutrino is impossible. The experimental evidence presented in this chapter suggests that the rest mass of the neutrino may be zero and therefore appears to agree with the predictions of this theory.

REFERENCES

1. Bethe, H. A. *Elementary Nuclear Theory*, Chapter VI. Wiley, New York, 1947.
2. Crane, H. R. *Rev. Mod. Phys. 20*, 278 (1948).
3. Haxby, R. O., Shoupp, W. E., Stephens, W. E., and Wells, W. H. *Phys. Rev. 58*, 1035 (1940).
4. Lyman, E. M. *Phys. Rev. 55*, 234 (1939).
5. Hughes, D. J., and Eggler, C. *Phys. Rev. 73*, 809 (1948); 1242 (1948).
6. Kofoed-Hansen, O. *Phys. Rev. 71*, 451 (1947).
7. Pruett, J. R. *Phys. Rev. 73*, 1219 (1948).
8. Hanna, G. C., and Pontecorvo, B. *Phys. Rev. 75*, 984 (1949).
9. Curran, S. C., Angus, J., and Cockroft, A. L. *Phil. Mag. 40*, 53 (1949); *Phys. Rev. 76*, 853 (1949).
10. Langer, L. M., and Moffat, R. J. D. *Phys. Rev. 88*, 689 (1952).
11. Hamilton, D. R., Alford, W. P., and Gross, L. *Phys. Rev. 92*, 1521 (1953).
12. Cook, C. S., Langer, L. M., and Price, H. C., Jr. *Phys. Rev. 73*, 1395 (1948).
13. Lee, T. D., and Yang, C. N. *Phys. Rev. 104*, 254 (1956); *Phys. Rev. 105*, 1671 (1957).
14. Lee, T. D., Oehme, R., and Yang, C. N. *Phys. Rev. 106*, 340 (1957).
15. Pauli, W. *Handbuch der Physik*, Vol. 24, pp. 226–227. Julius Springer, Berlin, 1933.

CHAPTER 3

Neutrino Recoils Following the Capture of Orbital Electrons

3.1. The Orbital Electron Capture Process

According to our present model of the orbital electron capture process, in an allowed transition an electron from the K shell, or less frequently from the L shell, is captured by the nucleus according to the scheme:

$$(Z + 1)^A + e^-_{K,L} \rightarrow Z^A + \nu, \tag{3.1}$$

where Z^A represents a nucleus of atomic number Z and mass number A. The conservation of energy in this decay process is given by:

$$W_\nu = M(Z + 1) - M(Z) - E_r - B_{K,L}, \tag{3.2}$$

where W_ν is the neutrino energy including the rest mass, $M(Z + 1)$ and $M(Z)$ are, respectively, the atomic masses of the initial and final atoms, E_r is the kinetic energy of the recoiling atom, and $B_{K,L}$ is the binding energy of either the K or L shell, all in the same energy units. If the single neutrino picture is correct, the energy spectrum of the recoiling atoms will consist of a sharp "line," since the entire energy of the transformation is carried off by the neutrino and the recoiling atom. In contrast, the emission of multiple neutrinos would give a continuous distribution of recoil energies. Since $W_\nu \gg (E_r + B_{K,L})$ in most neutrino experiments, the neutrino energy is given very closely by:

$$W_\nu = M(Z + 1) - M(Z). \tag{3.3}$$

If a single neutrino is emitted, the kinetic energy of the recoiling atom is given by:

$$E_r = 140.2 \frac{(W_\nu^2 - \nu^2)}{M} \text{ electron volts}, \tag{3.4}$$

where W_ν is in units of mc^2, ν is the neutrino mass in units of m, and M is the atomic mass of the recoiling atom. In practically all electron capture decays E_r is less than 100 ev. In principle the rest mass of the neutrino can be deduced from equations (3.2) and (3.4). The mass difference between the initial and final nuclei can be obtained from the threshold of the p, n reaction, which produces the initial nucleus together with the n-H mass difference. In order to set an upper limit of one percent of the electron mass for ν, both W_ν and E_r must be known to an

20

accuracy of 10^{-3} percent or less. This would require an improvement in the present accuracy of the values of E_r and W_r by a factor of 10^2 to 10^3. Apparently the values of the neutrino mass obtained from the low energy beta-decay experiments of Chapter 2 are considerably more accurate than those which can be deduced from electron capture experiments.

Studies of the neutrino recoils resulting from electron capture in A^{37} have yielded specific information regarding the neutrino, and have given us a new insight into the behavior of the various electronic shells of the atom following the removal of a vacancy in one of these shells. This latter effect is manifested by the appearance of multiply-charged positive recoil ions. For example, the average number of electronic charges per recoil ion is 3.0 ± 0.2 according to the measurements of Wexler,[1] and is 3.41 ± 0.14 according to the work of Perlman and Miskel.[2] The theoretical investigations of Rubenstein and Snyder[19] have shown that the shape of the charge distribution of the recoil ions can be explained by the emission of a cascade of Auger electrons as the initial K or L_1 vacancies are filled. This process will be described in greater detail in the section dealing with the A^{37} experiments.

A list of neutrino experiments based on the orbital electron capture decay process is given in Table 3.1. In each of these experiments an attempt has been made to measure the energy spectrum of the recoiling nuclei.

TABLE 3.1

Orbital Electron Capture Neutrino Experiments

Initial Nucleus	References
Be⁷	Allen[7]; Smith and Allen[8]; Davis[9]
A³⁷	Rodeback and Allen[11]; Kofoed-Hansen[12]; Snell and Pleasonton[13]; Rubenstein and Snyder[19]
Cd¹⁰⁷	Alvarez, Helmholz, and Wright[16]; Wright[17]

3.2. Neutrino Recoils Following Electron Capture in Be⁷

The nuclide Be⁷ decays exclusively by electron capture to Li⁷ according to the nuclear reactions:

$$\mathrm{Be}^7 + e^-_{K,L} \rightarrow \mathrm{Li}^7 + \nu,$$

$$\mathrm{Be}^7 + e^-_{K,L} \rightarrow \mathrm{Li}^{7*} + \nu, \tag{3.5}$$

$$\mathrm{Li}^{7*} \rightarrow \mathrm{Li}^7 + h\nu(478.5 \pm 0.5 \text{ kev}), \qquad \text{(reference 3)}$$

with $T_{1/2} = 53.6$ days. The Li^7 is left in an excited state in about ten percent of the disintegrations. A Be^7-Li^7 mass difference of 0.864 ± 0.003 Mev can be computed from the Q value (references 4 and 5) of 1.646 ± 0.002 Mev for the reaction Li$^7(p, n)$ Be^7 and the n-H mass difference[6] of 0.782 ± 0.001 Mev . According to equation (3.3), the total energy of the neutrino or neutrinos emitted in the first of reactions (3.5) is very closely equal to 0.864 Mev . Likewise, the total neutrino energy in the second reaction is 0.385 Mev. In ninety percent of the decays we should expect mono-energetic nuclear recoils with 57.3 ± 0.5 ev energy if a single neutrino is emitted. The ten percent branch going to the excited state of Li^7 should exhibit a continuous recoil spectrum extending from nearly zero to 57.3 ± 0.5 ev and resulting from the nearly simultaneous emission of a neutrino and a gamma ray.

In all of the Be^7 neutrino experiments carried out so far, the source has been a nearly monatomic layer of Be or perhaps BeO deposited on a suitable substratum. Since the basic orbital electron capture process does not in itself result in a charged nuclear recoil, the work function of the substratum must be greater than the first ionization potential of the recoiling atom in order for surface ionization to occur. Although surface ionization may play an important role in the production of charged recoil atoms, it is likely that ionization as a result of the Auger effect may also be of importance.

The first successful neutrino experiment was carried out by Allen.[7] The Be^7 used in this experiment was produced by the Li$^6(d, n)$ Be^7 reaction with LiF as the target material. Preliminary investigations showed that most of the Be^7 could be separated from the target material by selective vacuum evaporation. The final source for the recoil experiment consisted of a platinum foil bearing a thin layer of Be^7. The actual chemical composition of the Be^7 was somewhat uncertain, but the evaporation characteristics suggested that it was largely metallic Be.

The experimental arrangement used by Allen is shown in Fig. (3.1). The Be^7 source at S was mounted on a Pt strip which could be heated by means of an electrical current. The positive recoil ions passed through two grids and then were accelerated into an electron multiplier tube by a potential difference of 3.6 kv. The lower grid G_1 was supported by insulators and was used for retarding potential measurements of the energy spectrum of the recoil ions. The results of this experiment are shown in Fig. (3.2). There is clear evidence of a spectrum of neutrino recoils with a maximum energy of perhaps 48 ev. Although the expected energy was 57 ev, the most likely explanation was that the line spectrum of the recoils resulting from single neutrino emission had been degraded by absorption either in the source itself or in material on the surface of the source. A separate experiment, in which the number of gamma ray counts in coincidence with recoil counts was measured, indicated

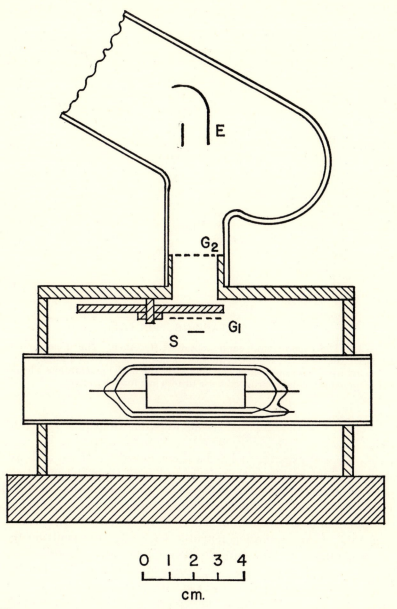

Fig. 3.1. The experimental arrangement used by Allen[7] in his Be⁷ neutrino experiment.

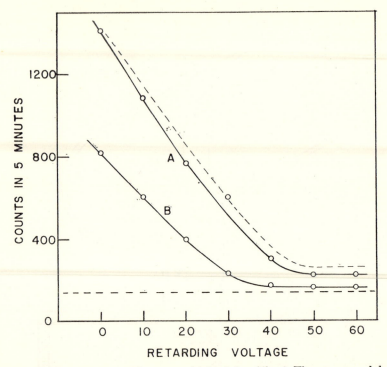

Fig. 3.2. The experimental curves obtained by Allen.[7] The upper and lower solid curves, respectively, represent the spectra of the recoils from a freshly prepared source, and from this same source one hour after the initial preparation. The dotted curve represents the data corrected back to zero age for the source.

that very few of the observed recoils were due to gamma ray emission alone.

A second Be[7] experiment has been conducted by Smith and Allen,[8] using apparatus similar to that shown in Fig. (3.1). In this investigation improved techniques were available for the preparation of the Be[7] source, and a new retarding grid structure permitted a more precise determination of the maximum energy of the recoils.

Fig. (3.3) shows retarding potential curves for the neutrino recoils from three different Be[7] sources. These sources were deposited on Ta foils by vacuum evaporation and were held at about 500°C during the experiments. The absolute source strength of each of the three sources was determined by counting the gamma rays emitted in the ten percent branch of the Be[7] decay scheme. The source efficiencies computed from the measured source strengths ranged from one percent for surface A to ninety-three percent for surface C. The high surface efficiency of surface C indicated that essentially all the recoils were able to leave the surface. Although the shape of curve C showed that relatively more high

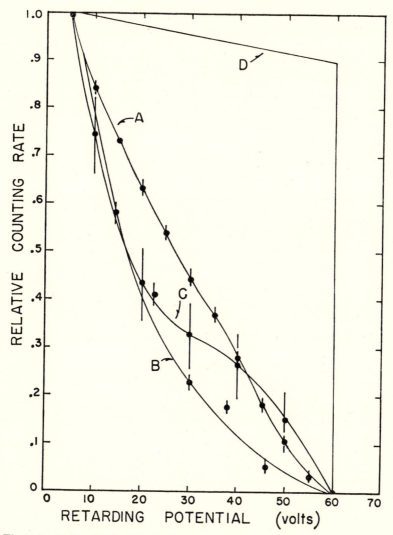

Fig. 3.3. *A*, *B*, and *C* are retarding potential curves for the neutrino recoils from Be⁷ surfaces of decreasing thickness. *D* is the curve expected if single neutrinos are emitted during the decay of Be⁷. From Smith and Allen.[8]

energy recoils were emitted from this source, the spectrum still did not agree with that corresponding to the emission of single neutrinos. However, the end point of the spectrum was at 56.6 ± 1.0 ev, in good agreement with the expected value of 57.3 ± 0.5 ev.

A third Be⁷ neutrino experiment has been carried out by Davis.[9] This experiment differed from the previous two investigations in the method used to measure the energy of the recoils. Davis used an electrostatic analyzer of relatively high resolution: $\Delta E_r/E_r = 0.018$. The differential energy spectrum of the neutrino recoils was measured rather than the integral spectrum as obtained by the retarding potential method.

A schematic diagram of the apparatus used by Davis is shown in Fig. (3.4). As in the earlier Be[7] experiments, an electron multiplier

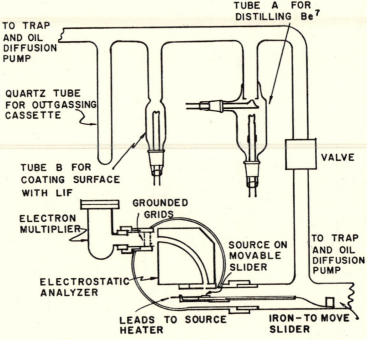

Fig. 3.4. The apparatus used by Davis[9] to study the recoil spectrum of Be[7].

was used to count the positively charged recoil ions. In order to increase the counting efficiency of the ion detector, the low-energy ions leaving the analyzer were accelerated by 2000 volts onto the first dynode of the electron multiplier.

The recoil energy spectra from a doubly distilled Be[7] source on a lithium fluoride surface are shown in Fig. (3.5). These curves were recorded with the surface temperature of the source at the indicated values, starting with room temperature. The curve of Fig. (3.6) represents the energy spectrum from a source on a tungsten ribbon. Most of the surface impurities had been removed before the measurements by flashing the ribbon at 1560°C and later at 1000°C. The temperature of the ribbon was held at 300°C for 3.5 hours, and then the data for Fig. (3.6) were recorded.

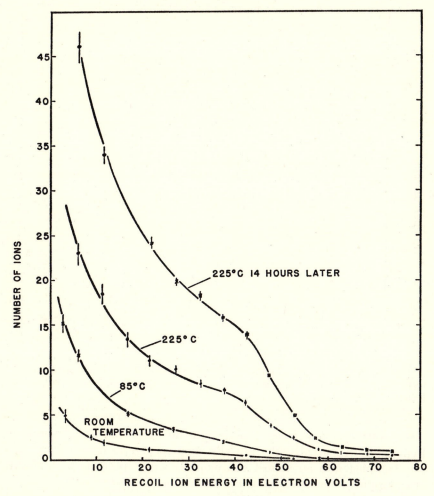

Fig. 3.5. The recoil spectra from a doubly distilled Be⁷ source on a lithium fluoride surface measured at a series of surface temperatures. From Davis.[9]

The curves of Figs. (3.5) and (3.6) show that the shape of the spectrum of the recoils from Be⁷ is strongly influenced by the nature of the sub-stratum of the source and also by the heat treatment of the source. There is some indication of a peak centered at about 48 ev in Fig. (3.6), indicating that the minimum energy loss experienced by the ions leaving the source was about 9 ev. The observed end point of the spectrum was 55.9 ± 1.0 ev, to be compared with the expected value of 57.3 ± 0.5 ev.

From this review of the Be⁷ neutrino recoil experiments it is evident that the experimental results do not clearly indicate the emission of single neutrinos in the orbital electron capture type of decay. However,

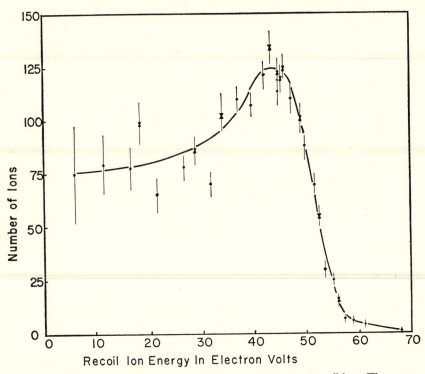

Fig. 3.6. The recoil spectrum from a Be⁷ source on a tungsten ribbon. The source had been held at a temperature of 300°C for 3.5 hours before these data were recorded. From Davis.[9]

the results of other recoil experiments do favor the single neutrino hypothesis. Apparently the initially mono-energetic recoils from Be⁷ are degraded into an almost continuous energy spectrum as a result of energy losses experienced at the surface of the source.

3.3. Neutrino Recoils Following Electron Capture in A³⁷

Argon-37 is an almost ideal substance for neutrino recoil experiments because it possesses a number of favorable properties. The predominant mode of decay is by electron capture. Since no gamma emission has been observed in this transition, the majority of the recoils should result from neutrino emission. Since argon is a monatomic gas, surface effects and disturbing molecular effects should be absent.

Assuming the emission of a single neutrino, the disintegration of A³⁷ by orbital electron capture is represented by:

$$A^{37} + e^-_{K, L} \rightarrow Cl^{37} + \nu, \tag{3.6}$$

with $T_{1/2} = 34$ days. The energy to be carried away by the neutrino is given by the A^{37}-Cl^{37} mass difference. A value of 816 ± 4 kev for this mass difference has been obtained from the measurement[10] of the threshold of the $Cl^{37}(p, n)A^{37}$ reaction, together with an n-H mass difference[6] of 782 ± 1 kev. The energy distribution of the recoils should be a sharp line centered at 9.67 ± 0.08 ev, if practically mono-energetic neutrinos are emitted. The width of this line is expected to be increased by several percent as a result of the added recoil due to the emission of Auger electrons and by a similar amount due to the thermal motion of the argon atoms.

The primary electron capture process in A^{37} should form electrically neutral Cl^{37} atoms. However, as a result of the emission of one or more Auger electrons approximately 10^{-10} seconds after the decay, most of the recoiling atoms are expected to be multiply-ionized. The maximum energy of the Auger electrons is about 2500 ev.

Rodeback and Allen[11] were the first to carry out a neutrino experiment using A^{37}. By observing the delayed coincidences between the Auger electrons and the corresponding recoil ions, the time of flight spectrum of the recoils was measured. A schematic of the A^{37} apparatus is shown in Fig. (3.7). Electron multiplier tubes were used as the detectors for

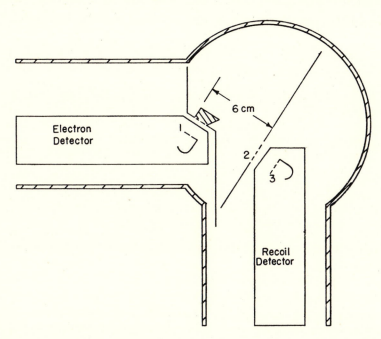

Fig. 3.7. Schematic of the A^{37} apparatus used by Rodeback and Allen.[11] The effective source volume is indicated by the shaded area in front of grid 1. The recoil Cl^{37} ions resulting from a disintegration within the source volume traverse a field-free path to grid 2 and then enter the ion counter after an acceleration through a potential difference of 4.5 kv.

both electrons and recoil ions. The shaded area represents a cross section of the effective source volume seen by both detectors. A delayed coincidence was recorded for an A^{37} disintegration occurring within the source volume when the resulting Auger electron passed into the electron detector, and in addition, the recoil ion entered the recoil detector. The A^{37} gas was not confined by differential pumping, but instead filled the entire recoil chamber at a pressure less than 10^{-5} mm of Hg. The radioactive gas was continuously circulated through the vacuum system and pumps during the experiment. Most of the impurities which accumulated were effectively removed in a hot calcium purifier. However, the pressure within the system slowly increased due to the accumulation of atmospheric argon which had entered through leaks. Despite this increase in pressure, measurements were made over periods of twenty-four hours.

The time-of-flight distribution of the recoils from A^{37} is shown in Fig. (3.8). Because of the relatively poor geometry, the peak expected for singly-charged Cl^{37} ions resulting from the emission of single neutrinos was a peak approximately 3 μsec wide at the base. This ideal time-of-flight distribution was modified by the effect of the thermal velocities of the A^{37} atoms in the source. Computations indicated that

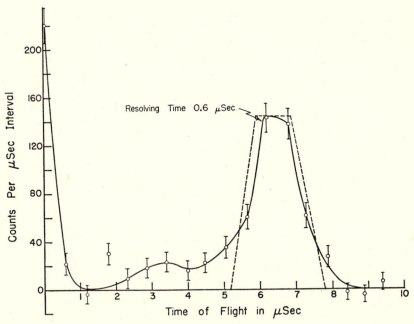

Fig. 3.8. The time of flight distribution of the neutrino recoils from the decay of A^{37}. The dashed curve is the distribution expected for monoenergetic recoils coming from the source volume. The tail of the solid curve in the region of 8 μsec is due to the thermal velocities of the A^{37} atoms in the gaseous source. From Rodeback and Allen.[11]

the shape of the distribution in the region of 8 μsec was fully explained by the thermal motion of the argon atoms. The intercept of the dotted curve at 7.8 μsec was assumed to represent the time of flight corresponding to the maximum path for the recoils. In order to compute the recoil velocity, it was necessary to take into account the time for the ions to traverse the path between grid 2 and the first dynode of the multiplier tube, and also the effect of the penetration of the electric field into the region between grid 2 and the source volume. The experimentally determined velocity was 0.71 ± 0.06 cm/μsec, which was in excellent agreement with the expected value of 0.711 ± 0.004 cm/μsec .

A search was made for possible low-energy recoils by extending the time-of-flight spectrum to 35 μsec . This measurement demonstrated the complete absence of low-energy recoils, at least down to 0.3 ev. The origin of the short time coincidences shown in Fig. (3.8) is not entirely understood. However, it appears likely that the broad peak in the region of 3.5 μsec was caused by multiply-charged recoil ions. Although all the recoils presumably had very nearly the same initial velocities, the accelerating field between grids 2 and 3 and also the leakage field to the left of grid 2 produced an increase in the final velocity, depending on the charge state of the ions. The peak at zero time probably was caused by coincidences between Auger electrons and x-ray photons, and also by coincidences between recoil ions and electrons resulting from disintegrations occurring between grids 2 and 3. In this latter case the electrons would be accelerated towards the source volume and then would be scattered into the electron counter. If we accept the above explanations for the origin of the peaks at zero time and at 3.5 μsec, then the main peak at 6.5 μsec clearly demonstrated that single, mono-energetic neutrinos are emitted during the decay of A³⁷.

The charge distribution of the recoil ions from the decay of A³⁷ has been extensively studied by Kofoed-Hansen[12] and also by Snell and Pleasonton.[13] Kofoed-Hansen has tried to eliminate the effects of poor geometry present when gaseous sources are used. He has used a crossed field method with relatively simple geometry. The instrument consisted of two plane parallel condenser plates with a uniform magnetic field parallel to the surface of the plates. The A³⁷ gas at a low pressure filled the space between the condenser plates. The central part of one of the plates was used as a collector plate surrounded by a guard ring. The current of recoil ions and Auger electrons going to the collector was measured for various ratios of the electric and magnetic fields. Information about both the average kinetic energy and the charge spectrum of the recoil ions was deduced from the shape of the current plots. The measured value of the average energy of the recoils was 9.6 ± 0.2 ev, corresponding to a neutrino energy of 812 ± 8 kev. This result clearly indicates the presence of very nearly mono-energetic neutrinos, since the measured value of the average recoil energy is in good agreement with the value of 9.67 ± 0.08 ev expected if single neutrinos are emitted.

The work of Snell and Pleasonton[13] closely paralleled that of Kofoed-Hansen both in objectives and also in time, but with a different experi-

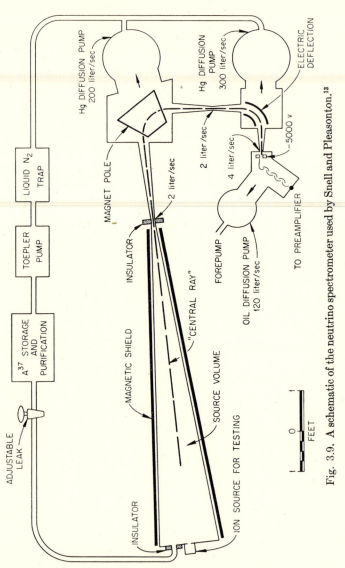

Fig. 3.9. A schematic of the neutrino spectrometer used by Snell and Pleasonton.[13]

mental approach. The recoil spectrometer of Snell and Pleasonton is shown in Fig. (3.9). A conical source volume contained the radioactive gas at a pressure of 2×10^{-5} mm of mercury or less. This volume was presumed to be field-free. About 10^{-6} of the recoiling atoms from disintegrations within the source volume emerged from a hole 0.5 inch in diameter at the small end of the cone. The resulting beam was subjected to analysis by deflection in a wedge-shaped stigmatic focusing magnetic field, followed by a second stigmatic focusing system with an electric field. Finally, the ions were accelerated through 4600 volts into an electron multiplier. The use of both magnetic and electric deflection resulted in the identification of the recoil ions both in terms of the charge-to-mass ratio and also according to energy.

The source volume was insulated from ground so that a predeflection acceleration could be applied to the ions before they entered the analyzer. When a positive bias of several hundred volts was applied to the source volume, the widths of the lines presented to the spectrometer became proportionately smaller and were exceeded by the transmission width of the spectrometer. At the other extreme, when the bias was not more than a few volts, the transmission width was smaller than the natural line width. In this second mode of operation the resolution of the spectrometer was sufficiently high to permit the determination of line shapes. For clarity we emphasize the fact that the lines referred to in this experiment were the groups of nearly mono-energetic recoil ions with charge states ranging from one to seven electronic units. In practice the line shapes were determined by matching the strength of the electric field to that of the magnetic field for a given charge-to-mass ratio. The line then was scanned by changing the accelerating voltage applied to the source volume.

A comparison of the natural shapes of the singly and triply charged recoil lines as observed by Snell and Pleasonton[13] is shown in Fig. (3.10). The continuous line gives the theoretical peak shape for mono-energetic recoils corrected for the thermal motion of the A³⁷ atoms, the *rms* value of the thermal velocity being about six percent of the neutrino recoil velocity. The dotted line shows the expected peak shape when the recoil resulting from the emission of randomly directed 2300 ev Auger electrons was added to the thermal effect. Although these curves were not intended for accurate recoil energy measurements, the shapes of both are consistent with the hypothesis of the emission of single neutrinos in the decay of A³⁷. Apparently the singly charged recoils were not accompanied by the emission of the 2300 ev Auger electrons. Snell and Pleasonton suggest that the following processes could lead to a singly charged recoil ion accompanied by an electron with energy too low to produce an appreciable spread of the line width:

(1) *K*-capture followed by fluorescent emission of the *K* x-ray. This is known to occur in about nine percent of the decays,[14] and will result in a single vacancy in the *L*-shell. The atom is still neutral, so one subsequent Auger process is called for.

(2) *L*-capture, known to occur in eight percent of the decays,[15]

resulting in a single *L*-vacancy, and requiring one subsequent Auger process to produce a singly charged ion.

Two determinations were made of the recoil energy using the peak corresponding to singly charged ions. The ions were accelerated to an energy of about 50 ev for the first measurement, and to about 90 ev for the second. The average value of the recoil energy was 9.63 ± 0.06 ev, to be compared with the value of 9.67 ± 0.08 ev based upon the

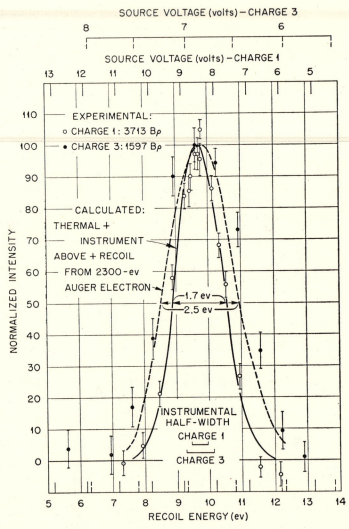

Fig. 3.10. A comparison of the natural shapes of the singly and triply charged recoil lines from A[37]. It is to be noted that the ion energies were very low and that the instrumental transmission width was less than the natural widths of the lines. From Snell and Pleasonton.[13]

$Cl^{37}(p, n)A^{37}$ threshold. Snell and Pleasonton prefer to use the value 9.65 ± 0.05 ev based on a later threshold measurement. In either case, however, the agreement between the experimentally determined value of the recoil energy and the value expected for single neutrinos of zero rest mass was extremely close. Unfortunately the accuracy of the energy determination is not sufficient for a precise determination of the rest mass of the neutrino. For example, from equation (3.4) the uncertainty of ± 0.05 ev in the recoil energy measurements could represent an uncertainty in the neutrino mass of about $\pm 0.1m$ or about ± 51 kev.

The distribution in charge of the neutrino recoils resulting from orbital

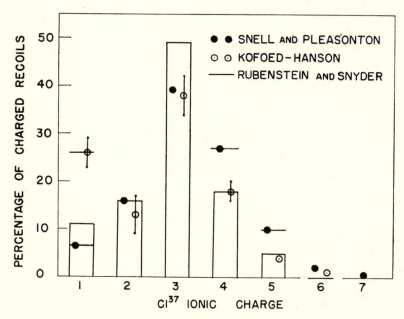

Fig. 3.11. The distribution in charge of the A³⁷ neutrino recoils. The histogram represents the distribution computed by Rubenstein and Snyder.[19]

electron capture in A^{37} is shown in Fig. (3.11). The experimental data are from the work of Kofoed-Hansen,[12] and Snell and Pleasonton.[13] The theoretical distribution indicated by the histogram was computed by Rubenstein and Snyder.[19] A Hartree self-consistent field method was utilized to obtain the electronic wave functions which were used in the evaluation of the matrix elements for the Auger and radioactive transitions. In order to explain the multiple ionization of the recoil atoms, Rubenstein and Snyder made the following assumptions:

(1) Almost every L_1 vacancy is filled by a Coster-Kronig[20] transition of the type $L_1 \rightarrow L_{11,111}M$, in which an L_1 vacancy is destroyed while

an M vacancy and either an L_{11} or an L_{111} vacancy is created with the ejection of an Auger electron. This type of transition occurs with a probability of unity.

(2) Every $L_{11,111}$ vacancy is filled by an Auger process since radiation rates are much smaller than Auger transition rates in the L shell.

(3) No mechanism exists whereby an M vacancy may be filled by an Auger process.

When the various Auger processes originating with a vacancy in the K shell were added to the list above, the production of positive recoil ions carrying up to five units of charge could be explained. A reasonably good agreement between the theoretical curve and the experimental results was obtained when it was assumed that A^{37} decayed by 0.92 K capture and 0.08 L_1 capture as reported by Pontecorvo, Kirkwood, and Hanna.[15] Perhaps the most obvious discrepancy between the experimental and theoretical results is the unexplained presence of ionic charges of $+6$ and $+7$ electronic units.

Snell has attempted to account for the emission of seven electrons by Auger processes, starting with a single K shell vacancy. He assumes additional Auger processes of the type $M_1 \rightarrow M_{11}M_{111}$ in order to fill M vacancies. Although it is doubtful as to whether this type of transition is energetically possible, Wolfsberg and Perlman[21] have suggested that there may be an electronic excitation due to the sudden change in the nuclear charge. A process of this type might excite the M electrons to the extent that the $M_1 \rightarrow M_{11}M_{111}$ transition is energetically possible. An alternative explanation is that two of the M electrons are independently ionized by this "shake-up" process.

3.4. Neutrino Recoils Following Orbital Electron Capture in Cd^{107}

One other orbital electron capture experiment has been reported. This investigation using 6.7 hr Cd^{107} was initiated by Alvarez, Helmholz, and Wright [16] and was continued by Wright.[17]

The decay scheme of Cd^{107} has been given by Bradt et al.,[18] and is reproduced in Fig. (3.12). According to this scheme, electron capture directly to a metastable level in Ag^{107} occurs in 99.27 percent of the disintegrations. The total energy available for the neutrino is 1.25 Mev, and a silver recoil ion of 7.9 ev should appear. In Wright's experiment a Cd^{107} surface was prepared by a double vacuum distillation. This active surface was exposed to a collector for a short time in order to collect the Ag^{107} recoil atoms which escaped from the surface of the source. Detection of the recoil atoms was accomplished by counting the conversion electrons from the 44 sec Ag^{107}*. By a comparison of the amount of 6.7 hr Cd activity on the source with the activity of the

Ag¹⁰⁷* on the collector, an efficiency of eight percent for the collection of the recoils was indicated. That is, of all the Ag¹⁰⁷ atoms recoiling into the solid angle available for collection, more than eight percent were actually collected. The fact that the collection efficiency was independent of the sign of the electric field between the source and collector indicated that the recoils were electrically neutral on leaving the surface of the source. Since the first ionization potential of Ag is 2 to 3 volts higher than the work function of the clean tungsten surface used as a sub-

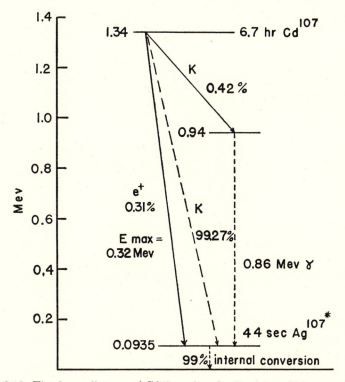

Fig. 3.12. The decay diagram of Cd¹⁰⁷ as given by Bradt[18] and his co-workers.

stratum, few, if any of the Ag recoils were ionized on leaving the surface of the source.

The fact that the collection efficiency was eight percent led Wright to conclude that the Ag¹⁰⁷ recoils resulted from neutrino emission during the electron capture process and not from the positrons or gamma rays produced in less than one percent of the transitions. However, since the energy distribution of the recoil atoms was not measured, this experiment did not yield specific information regarding the neutrino hypothesis.

In summarizing the results of the various electron capture neutrino experiments, it is evident that in the case of the Be^7 and Cd^{107} experiments there was no definite confirmation of the single neutrino hypothesis. However, recoil atoms from the decay of Be^7 were observed, and the maximum energy of these recoils agreed closely with the value predicted on the assumption that the missing energy and momentum were carried off by one or more neutrinos of essentially zero rest mass. The results of the A^{37} experiments clearly show that mono-energetic single neutrinos are emitted, and that the neutrino rest mass is not more than $0.1\ m$.

REFERENCES

1. Wexler, S. *Phys. Rev. 93*, 182 (1954).
2. Perlman, M. L., and Miskel, J. A. *Phys. Rev. 91*, 899 (1953).
3. Ter Pogossian, M., Robinson, J. E., and Goddard, C. H. *Phys. Rev. 76*, 1407 (1949).
4. Herb, R. G., Snowdon, S. C., and Sala, O. *Phys. Rev. 75*, 246 (1949).
5. Shoupp, W. E., Jennings, B., and Jones, W. *Phys. Rev. 76*, 502 (1949).
6. Taschek, R. F., Argo, H. V., Hemmendinger, A., and Jarvis, G. A. *Phys. Rev. 76*, 325 (1949).
7. Allen, J. S. *Phys. Rev. 61*, 692 (1942).
8. Smith, P. B., and Allen, J. S. *Phys. Rev. 81*, 381 (1951).
9. Davis, R. *Phys. Rev. 86*, 976 (1952).
10. Richards, H. T., Smith, R. V., and Browne, C. P. *Phys. Rev. 80*, 524 (1950).
11. Rodeback, G. W., and Allen, J. S. *Phys. Rev. 86*, 446 (1952).
12. Kofoed-Hansen, O. *Phys. Rev. 96*, 1045 (1954).
13. Snell, A. H., and Pleasonton, F. *Phys. Rev. 97*, 246 (1955); *100*, 1396 (1955).
14. Broyles, C. D., Thomas, D. A., and Haynes, S. K. *Phys. Rev. 89*, 715 (1953).
15. Pontecorvo, B., Kirkwood, D. H. W., and Hanna, G. C. *Phys. Rev. 75*, 982 (1949).
16. Alvarez, L. W., Helmholz, A. C., and Wright, B. T. *Phys. Rev. 60*, 160 (1941).
17. Wright, B. T., *Phys. Rev. 71*, 839 (1947).
18. Bradt, H., Gugelot, P. C., Huber, O., Medicus, H., Preiswerk, P., Scherrer, P., and Steffen, R. *Helv. Phys. Acta 19*, 218 (1946).
19. Rubenstein, R. A., and Snyder, J. N. *Phys. Rev. 99*, 189 (1955).
20. Burhop, E. H. S. *The Auger Effect*. Cambridge University Press, Cambridge, England, 1952.
21. Wolfsberg, M., and Perlman, M. L. *Phys. Rev. 99*, 1833 (1955).

CHAPTER 4

The Electron-Neutrino Angular Correlation in Beta-Decay

4.1. Introduction

A knowledge of the nature of the interactions responsible for beta-decay would help us in our understanding of the fundamental character of the beta-decay process and its relation to other nuclear phenomena. In recent years conclusive evidence concerning the forms of these interactions has been provided by studies of the shapes of certain forbidden beta-spectra, and also by studies of the angular correlation between the electron and neutrino emitted in beta-decay. Additional information regarding the relative strengths of these interactions has also been obtained by comparing the experimental ft values with the calculated nuclear matrix elements. This latter method has been most successful in the case of mirror transitions and in a few other favored transitions where the matrix elements can be estimated with the necessary accuracy. For similar reasons the theoretical interpretation of electron-neutrino angular correlation experiments has been less complicated in allowed beta-decay processes than in forbidden beta-decays. In the following parts of this chapter we shall first describe certain features of conventional beta-decay theory which pertain to angular correlation measurements; then we shall discuss certain modifications to this theory which are required by the non-conservation of parity, charge conjugation, and the possible noninvariance of time reversal.

4.2. Electron-Neutrino Angular Correlation in Allowed Transitions

The electron-neutrino angular correlation in nuclear beta-decay depends on the type of light-particle nucleon interaction assumed in the Fermi theory.[1] This was first pointed out by Bloch and Moller.[2] The angular correlation has been investigated theoretically by Hamilton,[3] deGroot and Tolhoek,[4] Greuling and Meeks,[14] and others.

According to the Dirac theory five independent relativistically invariant expressions can be chosen for the interaction Hamiltonian. These are usually called the scalar, vector, tensor, pseudo-vector (or axial vector), and pseudo-scalar interactions, denoted respectively by S, V, T, A, and P. The complete beta interaction can be expressed as

39

a linear combination of these five invariants if derivatives of wave functions are not included. The expression for the interaction energy can then be expressed as:

$$H_\beta = C_S H_S + C_V H_V + C_T H_T + C_A H_A + C_P H_P, \qquad (4.1)$$

where the coupling constants C_X represent the strengths of the various interactions. In conventional neutrino theory the C_X can be taken to be real. The specific form of H_β is given in the numerous reviews of beta-decay theory.[4,13]

The transition probability to a state in which the momenta p and q of the electron and neutrino are included within the solid angles $d\Omega_e$ and $d\Omega_\nu$, respectively, is given by:

$$P(W, \theta_{e\nu}) \, dW \, d\Omega_e \, d\Omega_\nu \qquad (4.2)$$
$$= (2\pi)^{-5} \Sigma_e \Sigma_\nu \, |\textstyle\int H_\beta \, d\tau|^2 \, pWq^2 \, dW \, d\Omega_e \, d\Omega_\nu$$

where W is the electron energy including the rest mass. All energies will be expressed in units of mc^2 and all momenta in units of mc. Σ_e indicates that a sum is to be taken over the two possible directions of the electron spin, and Σ_ν has the same significance for the two directions of the spin of the neutrino.

DeGroot and Tolhoek[4] have evaluated the transition probability for allowed transitions using the general interation energy of equation (4.1). In this computation the relativistic terms appearing in the operators of (4.1) were neglected with the exception of the term representing the P interaction which has only the relativistic form. The rest mass of the neutrino was assumed to be zero, and units were used in which $\hbar = m_e = c = 1$. The transition probability in reciprocal time units of $mc^2/\hbar$ is:

$$P_\pi(W, \theta_{e\nu}) \, dW \, d\Omega_e$$

$$= \frac{1}{8\pi^4} F(\pm Z, W) pW (W_0 - W)^2 [(C_S^2 + C_V^2) \, |\langle 1 \rangle|^2 \qquad (4.3)$$
$$+ (C_T^2 + C_A^2) \, |\langle \sigma \rangle|^2 + C_P^2 \, |\langle \beta \gamma_5 \rangle|^2]$$
$$\times \left[1 + \frac{b}{W} + \frac{\lambda p}{W} \cos \theta_{e\nu} \right] dW \, d\Omega_e,$$

where W_0 is the maximum beta energy including the rest energy. $F(\pm Z, W)$ is the Coulomb correction factor which reduces to unity when the effect of the nuclear charge on the electron can be neglected. $F(+Z, W)$ applies to negative electron emission and gives greater relative weight to slow electrons due to their retardation by the Coulomb field. $F(-Z, W)$ refers to positron emission and depresses the low-energy end of the beta-spectrum. The lower signs in (4.3) refer to positron and the upper to electron emission. The constants b and λ are given by:

$$b_\mp = \pm 2\gamma \frac{C_V C_S |\langle 1 \rangle|^2 + C_T C_A |\langle \sigma \rangle|^2}{(C_V^2 + C_S^2) |\langle 1 \rangle|^2 + (C_T^2 + C_A^2)|\langle \sigma \rangle|^2 + C_P^2 |\langle \beta \gamma_5 \rangle|^2} ; \quad (4.4)$$

$$\lambda = \frac{(C_V^2 - C_S^2) |\langle 1 \rangle|^2 + \frac{1}{3}(C_T^2 - C_A^2) |\langle \sigma \rangle|^2 - C_P^2 |\langle \beta \gamma_5 \rangle|^2}{(C_V^2 + C_S^2) |\langle 1 \rangle|^2 + (C_T^2 + C_A^2) |\langle \sigma \rangle|^2 + C_P^2 |\langle \beta \gamma_5 \rangle|^2}. \quad (4.5)$$

The symbols $\langle 1 \rangle$, $\langle \sigma \rangle$, and $\langle \beta \gamma_5 \rangle$, respectively, represent the nuclear matrix elements for the (S, V), (T, A), and P interactions, $\gamma = [1 - (\alpha Z)^2]^{1/2}$, and α is the fine structure constant.

The term b/W in (4.3) is commonly called a "Fierz interference" term[6], and according to (4.4) is different from zero if either $C_S C_V \neq 0$ or $C_T C_A \neq 0$. The existence of this interference term would require an additional energy-sensitive term $(1 \mp b/W)$ in the expression for the shape of an allowed beta-spectrum. This effect would cause the Fermi plots of allowed beta-spectra to show a systematic deviation from linearity. The largest deviations would occur in high-energy spectra. Mahmoud and Konopinski[7] have made a systematic study of the beta-spectra of N^{13}, S^{35} and Cu^{64}, and have estimated the largest value that the Fierz term could have and yet be consistent with the observed spectra. Their results indicate that $-0.08 < b < 0.24$. Davidson and Peaslee[8] studied a similar group of allowed spectra and arrived at the value $b < 0.08$ for the T and A interactions, and $b < 0.20$ for the S and V interactions. More recently Schwartzschild[9] has made a careful study of the shape of the He^6 beta-spectrum and concludes that $b < +0.1$ or $C_A/C_T < +0.05$. Sherr and Miller[5] have measured the ratio of positron emission to electron capture in the decay of Na^{22} and deduce $C_A/C_T = (-0.01 \pm 0.02)$. In general the high degree of linearity of the Fermi plots of allowed decays indicates that the Fierz terms are small, if not zero. In our discussion of electron-neutrino angular correlation experiments we shall assume that the Fierz terms in equation (4.3) can be neglected. When the Fierz terms are neglected, (4.3) shows an angular correlation of $(1 + \lambda p/W \cos \theta_{ev})$ between the directions of emission of the electron and of the neutrino. Formally this effect results when the matrix elements of H_β are summed over the spin directions of the electron and neutrino. In the relativistic treatment the light particles can be described by four component Dirac spinor wave functions φ_e and φ_ν. If we neglect the effect of the Coulomb field on the electron, we may use plane wave solutions of the Dirac equation. In the case of allowed transitions (order $l = 0$), we may take φ_e and φ_ν outside the integral of H_β. For example: in the case of the scalar interaction a transition resulting in the emission of a negative electron and an anti-neutrino will have a probability proportional to:

$$\sum_e \sum_\nu |\{\varphi_e^\dagger, \beta \varphi_\nu\}|^2, \quad (4.6)$$

where the operator β is a 4×4 Dirac matrix (see Bethe[10] or Schiff[11] for the explicit forms of the Dirac matrices). The evaluation of the sum (4.6) may be carried out by performing the indicated matrix multiplications for each of the four possible spin combinations. This sum may also be evaluated in terms of the traces of the products of the Dirac matrices.[12] The evaluation by either method results in an angular correlation term:

$$1 - \frac{p}{W} \cos \theta_{e\nu}. \tag{4.7}$$

The angular correlation terms corresponding to the non-relativistic parts of the T, A, V interactions and also the P interaction may be evaluated in a similar manner.

The angular correlation factors for each of the five interactions and also the appropriate nuclear matrix elements are listed in Table (4.1).

TABLE 4.1

Angular Correlation between Electron and Neutrino for Allowed Transitions

Interaction	Correlation Factor	Nuclear Matrix Elements	Change of Nuclear Parity and Total Angular Momentum
S	-1	$\langle 1 \rangle$	$\Delta J = 0$; no change of parity
V	$+1$	$\langle 1 \rangle$	$\Delta J = 0$; no change of parity
T	$+\frac{1}{3}$	$\langle \sigma \rangle$	$\Delta J = 0$; ± 1 no $0 \to 0$; no change of parity
A	$-\frac{1}{3}$	$\langle \sigma \rangle$	$\Delta J = 0$; ± 1, no $0 \to 0$; no change of parity
P	-1	$\langle \beta \gamma_5 \rangle$	$\Delta J = 0$; change of parity

The selection rules for the S, V and T, A interactions, respectively, are frequently referred to as Fermi and Gamow-Teller rules. Since there is no evidence for the P interaction in allowed beta-decays, we shall not consider this interaction in our discussion of angular correlation experiments based on allowed transitions.

If the neutrino mass is not zero, the operation described by equation (4.6) results in an additional term in the angular correlation factor of equation (4.3). This correction factor has the form:

$$\mp \frac{\nu}{w_\nu w},$$

where ν is the rest mass of the neutrino in units of m, w_ν is the total energy of the neutrino in units of mc^2, and w is the total energy of the electron. The negative sign applies if a Dirac anti-neutrino accompanies negative electron emission. The plus sign would apply to Fermi's original formulation of the theory in which the Dirac neutrino accompanied negative electron emission. If a Majorana neutrino (one having only positive energy states) were emitted in the beta-decay process, this correction would vanish. The explicit form of the correction probably was derived first by Pruett.[21] According to the evidence presented in Chapter 2, the rest mass of the neutrino is very small and may even be identically zero if the two-component neutrino theory proves to be correct. For this reason the correction factor depending on the neutrino mass usually is not included in the expression for the transition probability in beta-decay.

A practical electron-neutrino angular correlation experiment should yield a value of the angular correlation factor λ. From this experimental value of λ we may hope to deduce either the single interaction or the relative strengths of the two interactions responsible for the transition, Since the direction of the neutrino cannot be measured directly, it is necessary to infer this quantity from the momentum and energy conservation between the electron, neutrino, and the final recoiling nucleus. In general it will be necessary to transform the $P(W, \theta_{e\nu}, \lambda)$ distribution of (4.3) into a $P(W, \Phi_{e\nu}, \lambda)$ distribution, where $\Phi_{e\nu}$ now is the observable angle between the electron and the recoiling nucleus. However, it will be more convenient here to use the distribution $P''(K, \lambda)$ for the purpose of demonstrating the effect of the electron-neutrino angular correlation on the shape of the energy spectrum of the recoiling nuclei. This distribution function expresses the probability-per-unit-energy range for the emission of a recoiling nucleus with relative kinetic energy $K = E_r/E_0$. $P''(K, \lambda)$ is obtained by first expressing $P(W, \theta_{e\nu}, \lambda)$ in terms of the variables W, K, and λ, namely $P(W, K, \lambda)$. An integration

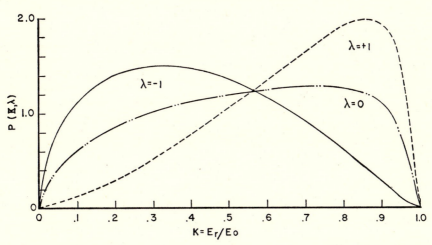

Fig. 4.1. The energy spectra for recoil nuclei emitted in an allowed beta-decay. The kinetic energy E_r of the recoil nucleus is expressed as a fraction of the maximum recoil energy E_0.

of $P(W, K, \lambda)$ for a fixed value of K, and over the entire range of W for which this value of K can occur will then yield a value of $P''(K, \lambda)$.

The recoil energy distributions predicted for angular correlations of $\lambda = \pm 1, 0$ are shown in Fig. (4.1). These curves were calculated with the Coulomb correction factor $F(Z, W) = 1.0$, which is exactly true only for $Z = 0$. However, only small changes in the shapes of the curves result when $Z < 20$ and $W_0 > 5mc^2$. It is evident that at both the high and low ends of the energy spectrum the shape of the spectrum is strongly influenced by the type of angular correlation used in the calculations. Because of this strong dependence, there is some hope that experimental investigations of the energy spectra of recoiling nuclei from allowed beta-decays will yield significant information regarding the form of the beta-decay interactions.

4.3. Electron-Neutrino Angular Correlation in Forbidden Transitions

The electron-neutrino angular correlation which appears in forbidden transitions usually is a rather complicated function of the nuclear matrix elements corresponding to several of the five beta-decay interactions. However, there is one particular group of forbidden transitions in which only a single matrix element corresponding to either the T or A interaction has an appreciable value. This circumstance makes predictions and interpretations particularly unambiguous, and is responsible for the name "unique" which is sometimes given to these transitions. These unique forbidden transitions occur when $|\Delta J| = n + 1$, where n is the degree of forbiddenness of the transition. The theory of this type of forbidden beta-decay has been reviewed by Konopinski,[13] and this source should be referred to for a description of the matrix elements and pertinent selection rules.

Hamilton[3] has derived the following expression for the electron-neutrino angular correlation in the case of a unique first forbidden beta-decay:

$$P(W, \theta_{e\nu}) = \text{const} \times \left[\left(5 \pm \frac{p \cos \theta_{e\nu}}{W} \right)(p^2 + q^2 + 2pq \cos \theta_{e\nu}) \right.$$
$$\left. \mp \frac{p^2 q \sin^2 \theta_{e\nu}}{W} \right], \quad (4.8)$$

where the upper and lower signs, respectively, are to be used for the T and A interactions.

Equation (4.8) was derived using plane waves to represent the electron, and therefore is correct only when $Z = 0$. However, Greuling and

Meeks[14] have arrived at the same angular correlation function when the effect of the nuclear Coulomb field is included. Both forms of equation (4.8) favor the emission of the electron-neutrino pair with $\theta \cong 0$. In general this will result in a recoil energy spectrum which is sharply peaked at the high-energy end. Several electron-neutrino angular correlation experiments based on unique first forbidden beta-decay processes will be described in Chapter 5. The results of these experiments will be compared with the predictions of equation (4.8).

4.4. Parity Conservation, Invariance under Time Reversal, and Charge Conjugation in Beta-Decay

At the end of Chapter 2 we introduced the subject of parity conservation in beta-decay. This general problem of the invariance under space inversion (parity), and also under time reversal and charge conjugation is so important in the theory of the neutrino that we shall discuss some additional implications here.

As Lee and Yang[15] have pointed out, the question may be raised as to whether the numerous experiments related to beta-decay have provided us with information concerning the conservation of parity in weak interactions. They have examined this question in detail and have demonstrated that the shapes of both allowed and forbidden spectra, the electron-neutrino, and electron gamma ray angular correlations have given no information concerning the conservation of parity. It should be emphasized here that these considerations of parity conservation do not refer to nuclear states. Apparently the usual selection rules for beta ray transitions still are valid.

Lee and Yang[15,A.1] have suggested an interaction Hamiltonian which contains, in addition to the five usual types of interactions, five new interactions which do not conserve parity. Time reversal invariance would require that all ten coupling constants C, C' be real, up to an overall phase. If this invariance does not hold, then beta-decay is characterized by ten complex coupling constants, i.e. twenty parameters. An additional restriction is that invariance under charge conjugation implies that C be real and that C' be purely imaginary, again up to an overall phase.

Lee and Yang[15] and others,[18] using the new interaction Hamiltonian mentioned in the previous paragraph have derived an expression for the energy and angular distribution of the electrons emitted in the allowed beta-decay of nonoriented nuclei. The general form of this expression is identical with that of equation (4.3). The Fierz interference term of equation (4.4) now is given by:

$$b\xi = \pm 2\gamma \operatorname{Re} \left[(C_S C_V^* + C_S' C_V'^*) |\langle 1 \rangle|^2 + (C_T C_A^* + C_T' C_A'^*) |\langle \sigma \rangle|^2 \right]; \quad (4.9)$$

and the angular correlation factor of equation (4.5) now is:

$$\lambda\xi = (|C_V|^2 + |C_V'|^2 - |C_S|^2 - |C_S'|^2) \, |\langle 1 \rangle|^2$$

$$+ \tfrac{1}{3} (|C_T|^2 + |C_T'|^2 - |C_A|^2 - |C_A'|^2) \, |\langle \sigma \rangle|^2$$

$$\pm 2 \, \mathrm{Re} \left\{ \frac{iZe^2}{\hbar c p} \, [(C_S C_V^* + C_S' C_V'^*) |\langle 1 \rangle|^2 \right. \tag{4.10}$$

$$\left. - \tfrac{1}{3}(C_T C_A^* + C_T' C_A'^*) \, |\langle \sigma \rangle|^2] \right\},$$

where:

$$\xi = (|C_S|^2 + |C_V|^2 + |C_S'|^2 + |C_V'|^2) \, |\langle 1 \rangle|^2 \tag{4.11}$$

$$+ (|C_T|^2 + |C_A|^2 + |C_T'|^2 + |C_A'|^2) \, |\langle \sigma \rangle|^2.$$

In the above expressions the unprimed and the primed constants, respectively, represent the relative strengths of the parity conserving and the parity non-conserving interactions. The upper signs refer to electron decay and the lower signs to the positron decay. It is evident that the angular correlation factor does not contain any interference terms between the parity conserving and parity non-conserving interactions. Therefore conventional electron-neutrino angular correlation experiments, i.e. experiments not carried out with polarized nuclei, or experiments in which the spin of electrons was not measured have not been able to separate the parity conserving from the parity non-conserving coupling constants. In other words, the conventional electron-neutrino experiment cannot show the degree of mixing of the C' type interactions with the usual type.

The third term of the angular correlation factor (4.10) does, however, show time reversal and also charge conjugation dependences. To be specific, this term will vanish if either time reversal is invariant or charge conjugation is invariant. Since the experiment of Wu et al.,[19] which will be described in Chapter 5 indicates that charge conjugation is violated, a definite p and/or Z dependence of the angular correlation factor λ would indicate that time reversal is non-invariant in beta-decay. However, if $|C_A|$, $|C_A'|$, $|C_V|$, and $|C_V'|$ are negligibly small compared to $|C_S|$, $|C_S'|$, $|C_T|$, and $|C_T'|$, the Coulomb correction to λ will be small, and this may make it difficult to detect possible violations of time reversal invariance by observation of the p and/or Z dependence of λ.

Before the discovery of parity violations in beta-decay, it was generally thought that the V and A coupling constants were small compared to the S and T coupling constants. However, the precise experiment of Sherr and Miller[5] shows only that the Gamow-Teller part of b is essentially zero, and according to equation (4.9) this proves only that Re $(C_T C_A^* + C_T' C_A'^*) = 0$. Winther and Kofoed-Hansen[20] have estimated that the Fermi part of b may be as large as 0.29. This value is rather uncertain because of the limited accuracy of the experimental data from which it was derived.

The reason for the absence of the interference terms CC' in the electron-neutrino angular correlation is rather obvious, and suggests the kind of experiment which could detect a possible interference. Since the angular correlation factor of equation (4.3) is proportional to a scalar $p \cdot q / W_e W_\nu$ which does not change sign under spacial inversions, the interference terms will not appear in the new expression for the angular correlation factor as shown by equation (4.10). However, interference terms could appear as a pseudoscalar, formed out of experimentally measured quantities. For example, if a momentum p and a spin σ are measured, the term $CC' \, p \cdot \sigma$ may occur. The product $p \cdot \sigma$ is a pseudoscalar and changes sign upon spacial inversion.

Lee and Yang[15] have suggested that a measurement of the angular distribution of the electrons coming from the beta-decay of oriented nuclei could establish whether parity conservation is violated in beta-decay. In the case of an allowed transition this angular distribution will involve terms of the form $\langle J_z \rangle \cdot p$, where $\langle J_z \rangle$ is the expectation value of the z component of the nuclear angular momentum, and p is the momentum of the electron. The angular distribution function can be represented by:

$$P(\theta) = \text{const} \times (1 + \alpha \cos \theta), \tag{4.12}$$

where θ now is the angle between the vectors J_z and p. If $\alpha \neq 0$, we would have a positive proof of parity non-conservation in beta-decay. In the special case of an allowed transition with $J \to J - 1$ (no change of nuclear parity), the quantity α is given by:

$$\alpha = \beta \frac{\langle J_z \rangle}{J},$$

with:

$$\beta = \text{Re} \left[(C_T C_T'^* - C_A C_A'^*) + i \frac{Ze^2}{\hbar c p} (C_A C_T'^* + C_A C_T'^*) \right]$$
$$\times |\langle \sigma \rangle|^2 \frac{p}{w} \frac{2}{\xi + (\xi b / W)}, \tag{4.13}$$

and

$$\xi = (|C_T|^2 + |C_T'|^2 + |C_A|^2 + |C_A'|^2) \, |\langle \sigma \rangle|^2. \tag{4.14}$$

A more general expression for the angular correlation has been given by Jackson et al.[18] According to Lee, Oehme, and Yang,[16] the first term in equation (4.13) vanishes if charge conjugation is invariant, and the second term vanishes if there is invariance under time reversal. The considerations above suggest that this type of experiment should be able to provide proof not only of a possible non-conservation of parity, but also of a possible non-conservation of charge conjugation. Since the second factor in equation (4.13) is reduced by the factor $Z/137p$, this term may be difficult to detect experimentally.

As mentioned at the end of Chapter 2, Lee and Yang[17] have proposed a two-component theory of the neutrino which is possible if parity is not conserved. They show that all calculations concerning beta-decay using the conventional theory of the neutrino with the Hamiltonian (A-1) of reference (15) give the same result as the two-component theory if the coupling constants are chosen as follows:

$$C_S = \mp C'_S, \qquad C_V = \mp C'_V, \quad \text{etc.} \tag{4.15}$$

The choice of the $\mp$ sign depends on whether the fundamental beta-decay process is:

$$n \rightarrow p + \bar{e} + \nu^* \text{(the anti-neutrino, a left-handed screw, } -), \tag{4.16}$$

or

$$n \rightarrow p + \bar{e} + \nu^* \text{(the anti-neutrino, a right-handed screw, } +). \tag{4.17}$$

If the two-component theory of the neutrino is applied to the problem of the angular distribution of the electrons coming from the beta-decay of oriented nuclei, then our equation (4.13) reduces to:

$$\beta = \mp \operatorname{Re} \frac{p}{W} \frac{|C_T|^2 - |C_A|^2 + i\dfrac{2Ze^2}{\hbar cp}(C_A C_T^*)}{|C_T|^2 + |C_A|^2}, \tag{4.18}$$

where the $\mp$ sign has the same significance as above. In writing equation (4.18) the Fierz interference terms have been set equal to zero in accordance with the experimental results discussed in Section 2 of this chapter. Since β must be a real quantity, equation (4.18) implies that the real part of $C_A C_T^* = 0$.

It is interesting to note that in positron emission the asymmetry parameter has a sign which is opposite to that of the parameter for electron emission. This is a direct consequence of the fact that in positron emission the neutrino emitted has opposite spirality from the anti-neutrino emitted in electron emission. According to equation (4.18) a measurement of the sign of the asymmetry factor should indicate the sign of the spirality of the neutrino or anti-neutrino emitted in beta-decay. In general, the relative magnitudes of $|C_T|^2$ and $|C_A|^2$ must be known if the interpretation of the measurement is to be unambiguous.

Wu et al.[19] recently have carried out an experiment similar to that suggested by Lee and Yang, and have found a definite asymmetry in the angular distribution of the electrons emitted in the beta-decay of oriented Co^{60} nuclei. This exceedingly important experiment will be discussed in greater detail in Chapter 5.

REFERENCES

1. Fermi, E. *Zeits. f. Physik 88*, 161 (1934).
2. Bloch, F., and Moller, C. *Nature 136*, 912 (1935).
3. Hamilton, D. R. *Phys. Rev. 71*, 456 (1947).
4. de Groot, S. R., and Tolhoek, H. A. *Physica 16*, 456 (1950).
5. Sherr, R., and Miller, R. H. *Phys. Rev. 93*, 1076 (1954).
6. Fierz, M. Z. *Physik 104*, 553 (1937).
7. Mahmoud, H. M., and Konopinski, E. J. *Phys. Rev. 88*, 1266 (1952).
8. Davidson, J. P., Jr., and Peaslee, D. C. *Phys. Rev. 91*, 1232 (1953).
9. Schwartzschild, A. Private communication (March 1956).
10. Bethe, H. A. *Handbuch der Physik*, Vol. 24, 1, Chapter 3. Julius Springer, Berlin, 1933.
11. Schiff, L. I. *Quantum Mechanics*, McGraw-Hill, New York, 1949.
12. Casimir, H. *Helv. Phys. Acta 6*, 287 (1933).
13. Konopinski, E. *Beta and Gamma-Ray Spectroscopy*, edited by Kai Siegbahn, Chapter X. North Holland Publishing Company, Amsterdam, 1955.
14. Greuling, E., and Meeks, M. L. *Phys. Rev. 82*, 531 (1951).
15. Lee, T. D., and Yang, C. N. *Phys. Rev. 104*, 254 (1956).
16. Lee, T. D., Oehme, R., and Yang, C. N. *Phys. Rev., 106*, 340 (1957).
17. Lee, T. D., and Yang, C. N. *Phys. Rev. 105*, 1671 (1957).
18. Morita & Morita, *Phys. Rev. 107*, 139 (1957); Jackson, et al. *Nuclear Phys. 4*, 206 (1957); Adler, et al. *Phys. Rev. 107*, 728 (1957); Morita, M. *Prog. Theoret. Phys. 10*, 345 (1953).
19. Wu, C. S., Ambler, E., Hayward, R. W., Hoppes, D. D., and Hudson, R. P. *Phys. Rev. 105*, 1413 (1957).
20. Winther, A., and Kofoed-Hansen, O. Private communication (1956).
21. Pruett, J. R. *Phys. Rev. 73*, 1219 (1948).

CHAPTER 5

Electron-Neutrino Angular Correlation Experiments

5.1. Introduction

The earliest recoil experiments were carried out in order to demonstrate that momentum was not conserved in radioactive disintegrations unless the existence of the neutrino was postulated. The experimental studies of the recoils from the orbital electron capture type of disintegration described in Chapter 3 have clearly demonstrated that the neutrino does carry off the missing momentum. Although the existence of an angular correlation between the directions of emission of the electron and the neutrino in beta decay processes requires that momentum and energy be carried off by the neutrino, the problem is complicated because of the presence of the electron. In general, recoil experiments on beta emitters have been carried out in order to obtain information regarding the form of the angular correlation, rather than to prove the conservation of momentum and energy. Angular correlation experiments using polarized nuclei, or experiments in which the polarization of the electron is measured should provide information concerning the conservation properties of parity, time reversal, and charge conjugation.

In beta-decay the maximum kinetic energy of the recoiling nucleus is given by:

$$E_0 = 140.2(W_0^2 - 1)/M \text{ ev}, \tag{5.1}$$

where M is the atomic mass of the final nucleus. This corresponds to a transition in which the momentum of the neutrino is essentially zero. For example, in the decay of He^6: $W_0 = 7.85 \ mc^2$, $E_0 = 1405$ ev; and in the decay of Ne^{19}: $W_0 = 5.27$, $E_0 = 205$ ev.

It is evident that in most recoil experiments low-energy negative or positive ions must be detected. An electron multiplier tube of the Allen[1] type with activated AgMg or BeCu surfaces has been used as the ion detector in recent experiments. In order to increase the number of secondary electrons produced by the impact of the ions upon the first electrode, the ions are accelerated through a potential difference of 1 to 10 kv just before entering the electron multiplier. Morrish and Allen[2] using this type of electron multiplier have shown that singly charged 2 kev lithium ions can be counted with an efficiency approaching

100 percent, and Robson[3] also has demonstrated that 4.2 kev protons can be detected with an efficiency of one hundred percent.

When the electron must be recorded in coincidence with the recoiling nucleus, the scintillation counter has proved to be the most convenient type of electron detector.

TABLE 5.1

Electron-Neutrino Angular Correlation Experiments

Initial Nucleus	log ft	References
n	3.09	Snell and Miller,[5] Snell, Pleasonton, and McCord,[6] Robson,[8] Spivac, Sosnovsky, Prokofiev, and Sokolov.[9]
He[6]	2.91	Allen, Paneth, and Morrish[13], Allen and Jentschke,[14] Rustad and Ruby[15]
Li[8]	5.60	Christy, Cohen, Fowler, Lauritsen, and Lauritsen[16]
C[11]	3.59	Leipunski[17]
Ne[19]	3.27	Alford and Hamilton,[46] Maxson, Allen, and Jentschke[20]
P[32]	7.90	Sherwin[25]
Cl[38]	7.44	Crane and Halpern[28]
A[35]	3.79	Herrmannsfeldt, Stahelin, Maxson, and Allen[58]
Kr[88]	7.7	Jacobsen and Kofoed-Hansen[32]
Kr[89]	6.19	Kofoed-Hansen and Kristensen[33]
Y[90]	7.98	Sherwin[34]

A list of electron-neutrino angular correlation experiments is given in Table (5.1). The methods used in these experiments and also the interpretation of the results will be discussed in the remainder of this chapter.

5.2. The Electron-Neutrino Angular Correlation in the Beta-Decay of the Neutron

Chadwick and Goldhaber[4] obtained an accurate value for the mass of the neutron and predicted that the neutron would undergo beta-decay into a proton, an electron, and a neutrino. The experimental confirmation of this prediction was first obtained by Snell and Miller,[5] who detected the presence of low-energy, positively charged particles resulting from the decay of thermal neutrons. In a later experiment Snell, Pleasonton, and McCord[6] obtained coincidences between the electrons and the low-energy protons emitted during the neutron decay. Robson[7] was able to measure the beta-spectrum by means of a magnetic spectrometer. A combination of a second magnetic spectrograph and an electrostatic field provided the means for the determination of the charge-to-mass ratio of the protons in coincidence with the electrons. The half-life of the neutron decay as measured by Robson

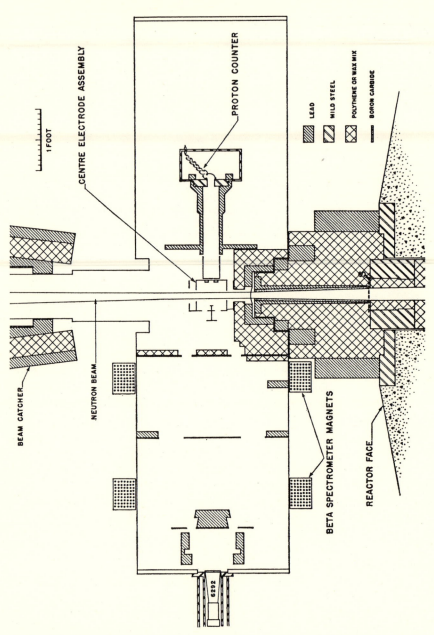

Fig. 5.1. Plan view of the apparatus used to study the neutron decay. From Robson.[8]

was 12.8 ± 2.4 minutes, and the maximum kinetic energy of the electrons was 782 ± 13 kev.

The beta-decay of the neutron can be represented by

$$n \rightarrow p + e^- + \nu, \qquad T_{1/2} = 12 \text{ min}, \qquad (5.2)$$

with $E_\beta^{\max} = 782 \pm 13$ kev, $E_0 = 760$ ev and a spin change of $J = \frac{1}{2}$ to $J = \frac{1}{2}$, no change of nuclear parity. According to Table (4.1) this decay is allowed by both Gamow-Teller and Fermi selection rules. Since this is a transition between mirror nuclei with true single particle states, exact values of the nuclear matrix elements can be computed. The values are $|\langle 1 \rangle|^2 = 1$ for the Fermi and $|\langle \sigma \rangle|^2 = 3$ for the Gamow-Teller type of decay.

In the experiment of Robson[8] the angular correlation was determined by measuring the momentum spectrum of those electrons which were emitted in coincidence with certain protons. The electrons were detected only if they were emitted in an angular range of approximately 145° to 175° from the direction of the proton beam. The angular spread of the proton beam was ±10° from the axis of the beta spectrometer. The recoil protons emitted in coincidence with the electrons were selected according to their transit times between the source volume and the proton counter.

Fig. (5.1) shows a plan view of the apparatus used by Robson. A collimated neutron beam from a reactor traversed an aluminum center-electrode assembly which was held at +7000 volts. The recoil protons leaving the source volume passed through a set of grids and were accelerated and electrostatically focused into the first dynode of an electron multiplier. The effective dimensions of the source volume were determined by the diameter of the beam of neutrons and also by the entrance slits of the beta spectrometer. The source determined in this manner was in the form of a cylinder with a diameter of about 2 cm and a length of 3 cm. Coincidences were sought between the electron counter and the proton counter, a time of flight analyzer being used to record the transit time of the protons. Fig. (5.2) shows the results of a typical run with the random coincidences subtracted. The peak occurred at the correct time and was of the correct width for protons resulting from neutron decays in the particular geometry used.

The momentum spectrum of the electrons was obtained by taking a series of proton-electron curves at various values of the momenta of the electrons. The result of a number of these runs is shown in Fig. (5.3). The best least squares fit was with an electron-neutrino angular correlation of $\lambda = +0.089 \pm 0.108$. This extremely small value of the angular correlation can be interpreted as an indication that either $|C_T| \cong |C_S|$ or that $|C_V| \cong |C_A|$. Since the results of the He^6 experiment clearly indicate that the Gamow-Teller interaction is C_T, we must

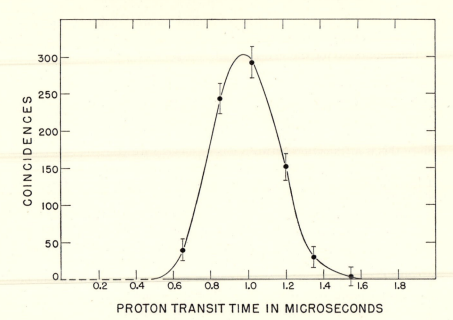

Fig. 5.2. Coincidences plotted against proton transit time. The beta spectrometer was set at 400 kev for this 24-hour run. From Robson.[8]

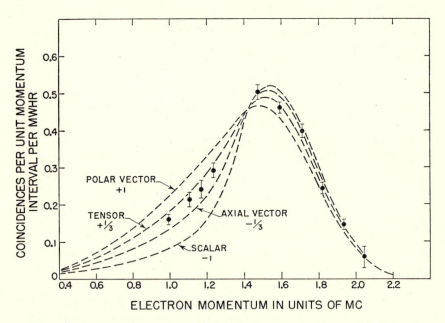

Fig. 5.3. The momentum spectra of the electrons in coincidence with the protons resulting from the decay of the neutron. The points represent the experimental data. From Robson.[8]

use the first choice. A substitution of Robson's value of λ in equation (4.5) gives $C_T^2/C_S^2 = 1.49\ ^{+1.44}_{-0.56}$ as the relative strengths of the Gamow-Teller and Fermi interactions. This result is quite unambiguous, since the value of the pseudoscalar matrix element $\langle \beta \lambda_5 \rangle$ which appears in the equation can be calculated. If we assume that the behaviour of both the neutron and the proton can be represented by a Dirac equation, we find that:

$$\langle \beta \lambda_5 \rangle^2 = \left(\frac{v_p}{c} \right)^2 \cong 10^{-6}. \tag{5.3}$$

This result rules out the possibility that the pseudoscalar type of coupling is partially responsible for the beta-decay of the neutron.

Spivac, Sosnovsky, Prokofiev, and Sokolov[9] have measured the half-life of the neutron using a method somewhat similar to that employed by Robson. In this experiment the half-life for the neutron decay was deduced from the rate of disintegrations occurring within a source volume containing thermal neutrons. The disintegration rate was measured by counting a known fraction of the recoil protons leaving the source volume. A value of 12 ± 1.5 minutes for the half-life of the beta-decay of the neutron was derived from these measurements.

A schematic diagram of the apparatus used by Spivac et al. is shown in Fig. (5.4). The beam of pile neutrons passed through the hollow, evacuated aluminum chamber A. A proportional counter C located near the center of A was supported on a long stem. A thin collodion window separated the counter from the vacuum inside A. A potential difference between electrode B and counter C was used to accelerate the protons into the proportional counter. Since the counting rate reached a constant value when the accelerating voltage exceeded 16 kv, the assumption was made that all of the protons were removed from the source volume. The absolute density of neutrons in the source volume was measured by the activation of a gold foil.

This same group has described an experimental study of the electron-neutrino angular correlation present in the decay of the neutron. The angular correlation was obtained from the shape of the time-of-flight spectrum of the recoil protons emitted in coincidence with the disintegration electrons. Their experimental arrangement consisted of three proportional counter proton detectors and three electron counters located in a plane perpendicular to the neutron beam. Each of the electron counters consisted of two Geiger-Müller counters in coincidence. The detectors were arranged to register simultaneously the time-of-flight spectrum of the protons at each of three different electron-proton angles. The results of a preliminary run are shown in Fig. (5.5). These preliminary data appear to favor the T interaction rather than the mixture of the T and S interactions observed by Robson.[8]

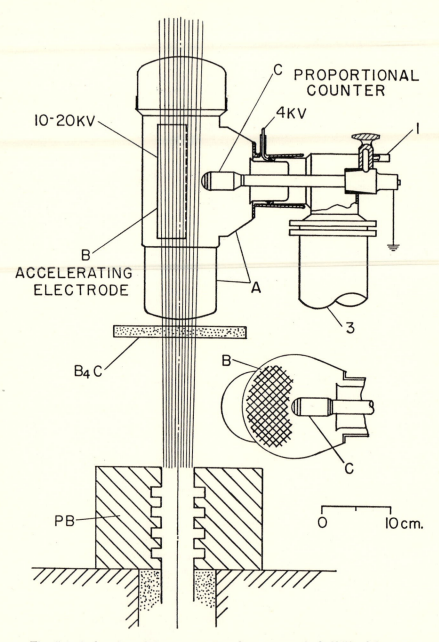

Fig. 5.4. A plan view of the apparatus used to measure the half-life of the neutron. Protons produced in the neutron beam were accelerated into the proton counter C by the electrostatic field in the region between electrode B and the counter. From Spivac, et al.[9]

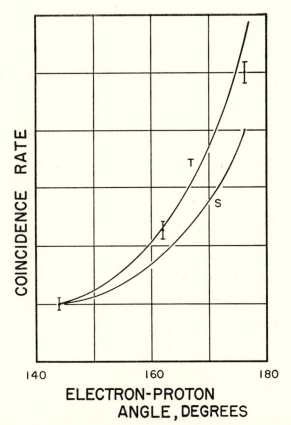

Fig. 5.5. The electron-neutrino angular correlation obtained by Spivac et al.[9] The theoretical curves have been computed for the T and S interactions.

5.3. The Angular Correlation in the Beta-Decay of He⁶

The beta-decay of He⁶ can be represented by:

$$\mathrm{He^6 \rightarrow Li^6} + e^- + \nu, \qquad T_{1/2} = 0.8\,\mathrm{sec}, \qquad (5.4)$$

with $E_\beta^{\max} = 3.50 \pm 0.05$ Mev,[10] $E_0 = 1405$ ev, and a spin change $J = 0$ to $J = 1$, no change of nuclear parity.

According to Table (4.1) this decay is forbidden according to the Fermi selection rules, but allowed according to Gamow-Teller selection rules, and consequently $|\langle 1 \rangle|^2 = 0$. The He⁶ transition occurs between the $^1S_0$ ground state of He⁶ and the $^3S_1$ ground state of Li⁶ which have nearly identical space wave functions in accordance with Wigner's theory of supermultiplets.[11] The value of the Gamow-Teller matrix element is $|\langle\sigma\rangle|^2 = 6$ if LS coupling is assumed for these nuclei. If we assume that the Fierz interference terms are absent, the electron-neutrino angular correlation factor will be either $\lambda = +\frac{1}{3}$ for the T interaction, or $\lambda = -\frac{1}{3}$ for the A interaction. The results of the He⁶ experiments are important in the theory of beta-decay since they indicate which of the two possible Gamow-Teller interactions is dominant in this transition.

Since the nuclear charge increases by one unit in the He⁶ decay, the recoiling Li⁶ atom is expected to carry a single positive charge. However, there is a possibility that ionization will accompany beta-decay due to the "shaking" of the atomic core as a result of the sudden change in the nuclear charge. Winther[12] has predicted that about ten percent of the Li⁶ ions appearing in the He⁶ decay will carry two units of positive charge. This presence of multiply-charged ions will be a possible source of error in certain types of recoil experiments unless the charge-to-mass ratio of the ions can be measured.

The first He⁶ experiment was carried out by Allen, Paneth, and Morrish.[13] In this experiment an attempt was made to obtain the recoil energy spectrum of the recoil ions in coincidence with electrons emitted at angles of 162° and 180° with respect to the direction of the ions. The He⁶ was produced by the reaction Be⁹(n, α)He⁶. Finely powdered beryllium metal was used as the target material and the radioactive gas was removed from the powder by a stream of ethyl alcohol vapor. The He⁶ gas was carried through a pipe line from the target to the recoil chamber. A cold trap was used to remove the alcohol vapor just before the He⁶ entered the vacuum system. A schematic diagram of the He⁶ apparatus is shown in Fig. (5.6). The He⁶ gas filled the entire chamber at a low pressure. A Geiger counter could be placed at opening A or B to count the electrons. The recoil ions were detected by the electron multiplier tube after an acceleration through a potential difference of 4.5 kv. A retarding potential method was used for the energy analysis of the recoil ions. The retarding voltage was applied to grid 4 located between grids 3 and 5. Delayed ion counts were recorded in coincidence with the electron counts.

The data obtained with the Geiger counter at position A are shown in Fig. (5.7). The effective source volume for the coincidences was determined by the intersection of the electron and recoil ion beams which were defined by the apertures shown in the diagram. It is evident that the energy of the recoil ions was about 1.4 kev, in good agreement with the predicted value. The statistical accuracy of the experimental data was so poor that the angular correlation could not be determined. However, the experimental points appear to lie within the $\lambda = \pm 1$ curves predicted for the Fermi type of transition, and probably are in better agreement with the $\lambda = -\frac{1}{3}$ curve of the Gamow-Teller theory. Considering the inaccuracy of the data, the retarding potential curves were not corrected for the presence of doubly charged recoil ions.

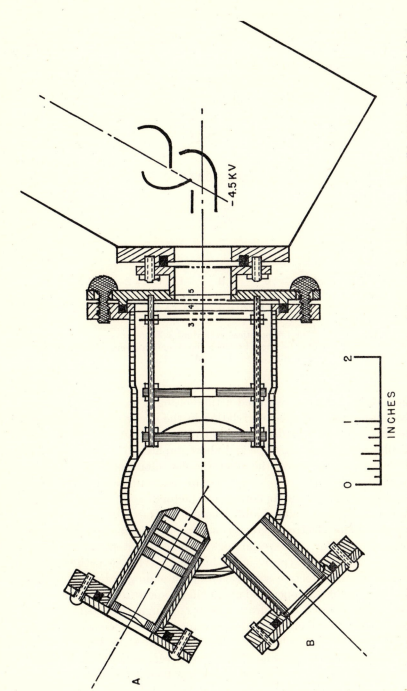

Fig. 5.6. An apparatus used for the retarding potential measurements of the energy spectrum of the recoil ions produced in the beta-decay of He⁶. The recoils were counted by the electron multiplier, and the electrons by a Geiger counter at window A. From Allen, Paneth and Morrish.[13]

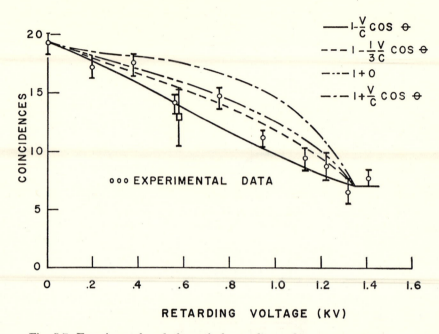

Fig. 5.7. Experimental and theoretical retarding voltage curves for the recoil ions from the decay of He[6]. The angle between the direction of emission of the electrons and the recoil ions was $162° \pm 8°$. From Allen, Paneth, and Morrish.[13]

Allen and Jentschke[14] have used a time-of-flight method to measure the electron-neutrino angular correlation in the decay of He[6]. Delayed coincidences between the disintegration electrons and the recoil ions were displayed on an oscilloscope and photographed on a slowly moving film. The data recorded on the photographic film were plotted as a time-of-flight spectrum which was compared with spectral curves computed for various angular correlation factors.

A diagram of the vacuum chamber used in this experiment is shown in Fig. (5.8). The electron detector was a scintillation counter with a stilbene scintillator. The recoil ions were counted by an electron multiplier of the same type as used in the first He[6] experiment. In order to define the effective source volume, a sliding gate A was employed. This gate had an open and also a closed aperture. The aluminum foil covering the closed aperture was thick enough to stop the recoil ions, but also was sufficiently thin to transmit the electrons without appreciable scattering or energy loss. The difference between the counting rates with the gate open and closed represented the net rate produced by disintegrations occurring in the region between the gate and stilbene scintillator. For most of the effective source volume the electron-recoil ion angle varied between $100°$ and $180°$.

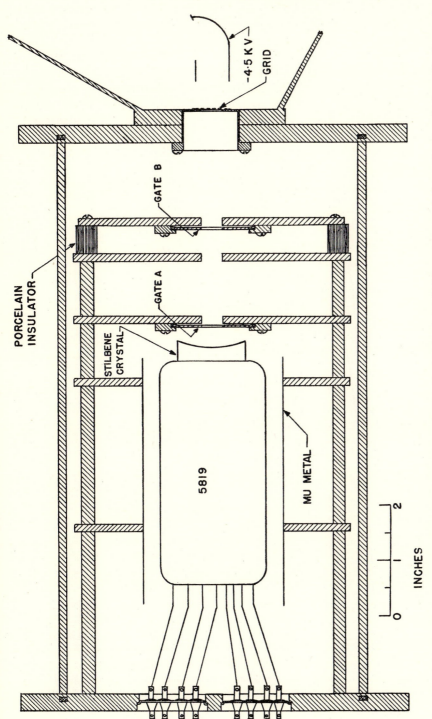

Fig. 5.8. The apparatus used by Allen and Jentschke (14) to obtain the time-of-flight spectrum of the nuclear recoils from the decay of He⁶. The effective source volume was cylindrical in shape and was defined at one end by the curved surface of the stilbene crystal and at the other by the gate A. The sliding gate B supported an alpha particle source used for testing the two detectors.

The experimentally determined time-of-flight spectrum is represented by the histogram of Fig. (5.9). In order to compute the curves expected for either the T or A interactions, the source was divided into three regions by planes parallel to the surface of the sliding gate. Recoil distributions were computed for a point source in the center of each of these regions. For each interaction the final distribution curve was a weighted mean of the individual curves. Although the statistical accuracy of the experimental data was poor, the observations were in better agreement with the curve representing the T interaction than with the curve for the A interaction.

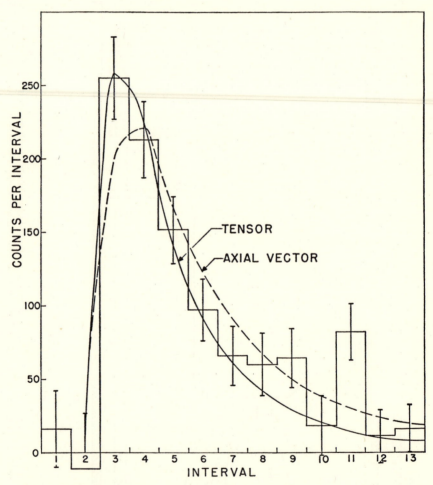

Fig. 5.9. The time-of-flight spectrum of the Li^6 recoil ions from the decay of He^6. The angle between the directions of emission of the ions and the electrons varied from about 100° to 180°. The width of a time channel was 6×10^{-8} sec. From Allen and Jentschke.[14]

The most comprehensive study of the electron-neutrino angular correlation in the decay of He⁶ has been conducted by Rustad and Ruby.[15] The angular correlation was deduced from the shape of distribution curves which represented the probability of an electron of energy W being emitted at an angle Φ with respect to the direction of the recoiling nucleus. Coincidences between the nuclear recoils and the electrons were recorded as a function of Φ and for essentially constant W in order to obtain the distribution curves.

The He⁶ was produced as in earlier experiments by the $Be^9(n, \alpha)He^6$ reaction. However, in this case the beryllium was in the form of $Be(OH)_2$ powder. The actual target consisted of 250 grams of this powder placed in an aluminum chamber located near the center of the Brookhaven reactor. Ethyl alcohol vapor was used to sweep the He⁶ gas from the bombardment chamber into the source volume. The apparatus used in this recoil experiment is illustrated by schematic sketches in Figs. (5.10) and (5.11). The essential elements were a defined source of He⁶, a collimator and an electron multiplier for detecting recoil nuclei, and a stilbene scintillation spectrometer to measure the direction and energy of the electrons.

The source volume as shown in Fig. (5.11) was defined by a thin aluminum foil and a series of differential pumping diaphragms. Both

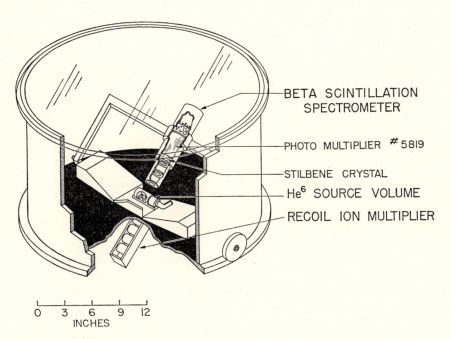

Fig. 5.10. Schematic sketch of the apparatus used to determine the electron-recoil ion correlation in the decay of He⁶. From Rustad and Ruby.[15]

the scintillation spectrometer and the source volume were enclosed in a bell jar which was evacuated in order to prevent the collapse of the thin covering of the source volume. As a result of the differential pumping, approximately ninety percent of the radioactive gas was removed through the gap separating the two sets of diaphragms. Counting measurements with the potential of the repelling grid, adjusted to shut off the ions coming from the source volume, indicated that the pressure of the He⁶ in the source volume was at least one hundred

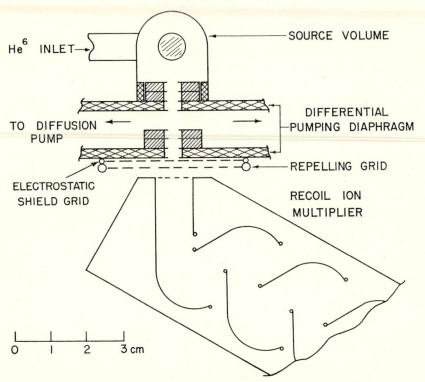

Fig. 5.11. A schematic diagram of the source volume, differential pumping diaphragms, and the ion detector used by Rustad and Ruby.[15]

times that in the multiplier region. As a result of the differential pumping, the effective source volume, although sharply defined by the aluminum foil, was defined somewhat less sharply by the diaphragms.

The results of the two electron recoil ion angular correlation measurements are shown in Figs. (5.12) and (5.13). In the first of these two measurements the scintillation counter was adjusted to accept electrons within the energy range 2.5 to 4.0 mc^2, and in the second the energy range from 4.5 to 5.5 mc^2 was used. It is evident that in both experiments the data points were very close to the curve expected for the T

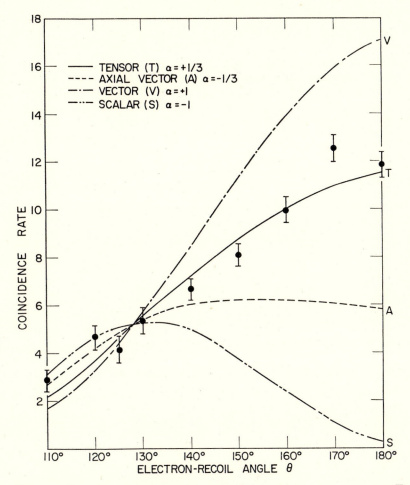

Fig. 5.12. The first electron-recoil ion angular correlation measurement. The scintillation spectrometer was adjusted to accept electrons in the energy range 2.5 to 4.0 mc^2. From Rustad and Ruby.[15]

interaction. A third set of measurements made near the upper end of the beta-spectrum was in good agreement with the curves obtained at lower electron energies.

Although the experimental results indicated that the beta-decay interaction for the He⁶ → Li⁶ transition is dominated by the tensor invariant, the data were further analyzed in an attempt to determine the possible presence of a small amount of the axial vector interaction. If we assume that the Fierz interference term is not zero, the angular correlation factor of equation (4.3) can now be expressed as:

$$(1 + b/W)(1 + \lambda'v/c \cos \theta_{ev}),$$

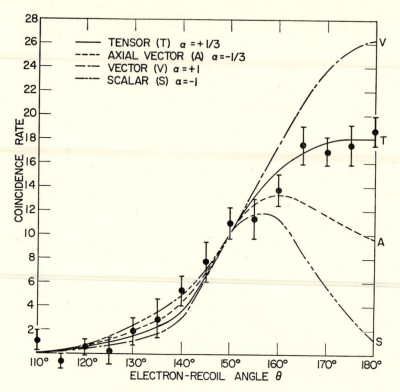

Fig. 5.13. The second electron-recoil ion angular correlation measurement. The energy of the electrons was in the range 4.5 to 5.5 mc^2.

with

$$\lambda' = \frac{(1 - x^2)W}{3[(1 + x^2)W + 2x]}, \qquad x = C_A/C_T. \tag{5.5}$$

The quantity b (pre-parity) has been defined in (4.4). In Fig. (5.14) the angular correlation factor λ' has been plotted as a function of X for a mean electron energy 2.0 Mev. The coefficients $\lambda_{1.24}$ and $\lambda_{2.0}$, respectively, were derived from the angular correlation curves corresponding to the electron energy ranges of 2.5 to 4.0 and 4.5 to 5.5 mc^2. The authors conclude that their results indicate that $-0.5 < C_A/C_T < +0.3$, which again shows that the tensor is the predominant interaction in the He^6 beta-decay. If we assume that the Fierz interference term is negligible, then the mean of their experimental values of the angular correlation would appear to be $\lambda = +0.34 \pm 0.09$. This result is to be compared with $\lambda = +\frac{1}{3}$ for a pure tensor interaction, and $\lambda = -\frac{1}{3}$ for a pure axial vector interaction.

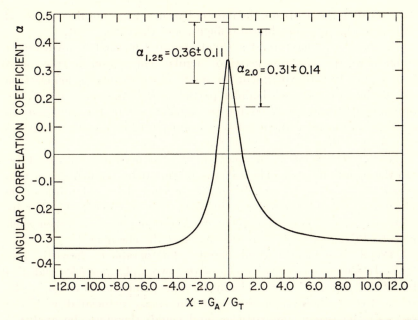

Fig. 5.14. The angular correlation coefficient $\lambda' = \frac{1}{3}(1 - x^2)W/(1 \times X^2)W + 2X$ plotted as a function of C_A/C_T. The values of λ' shown at the peak of the curve were determined from the electron-recoil ion correlation of He^6. From Rustad and Ruby.[15]

5.4. Nuclear Recoils from the Decay of Li⁸

Numerous investigators have proposed the study of the conservation of momentum in the decay of Li^8, which is unique in that it is followed by the break-up of the daughter nucleus Be^8 into two alpha particles. The decay scheme is:

$$Li^8 \rightarrow Be^8 + e^- + \nu$$
$$\searrow$$
$$He^4 + He^4, \qquad (5.6)$$

with the initial electron decay having a half-life of 0.88 second, followed by the alpha particle decay in a time of the order of 10^{-21} seconds. The Be^8 level is extremely broad, but the maximum probability is for the alpha pair to have about 3 Mev and the electron and neutrino together to have about 12 Mev. In principle, the Li^8 may be introduced into a cloud chamber and measurements made of the energy and momenta of the two alpha particles and of the electron. If the energy and momenta can be obtained with sufficient accuracy, the direction and momentum of the neutrino can be found by completing the momentum diagram. Although this experiment is capable of yielding rather specific information regarding the neutrino, the required accuracy has not been achieved because of the severe technical difficulties of the experiment.

Christy et al.[16] have used a cloud chamber method in an attempt to measure the electron-neutrino angular correlation in the beta-decay of Li^8. In this experiment the tracks of the disintegration electron and of the two associated alpha particles from the break-up of Be^8 were recorded and analyzed for the momentum of the neutrino. In principle, the form of the angular distribution of the electron-neutrino pair could be derived from these measurements.

A clever technique was employed to introduce the short-lived Li^8 into the cloud chamber. The Li^8 was produced by bombarding a thin layer of Li^7 with deuterons. Immediately after the bombardment the thin foil containing the Li^8 was introduced into a cloud chamber by means of a rapidly moving plunger. A photograph was taken each time the Li^8 source arrived at the center of the chamber.

A sketch of a typical cloud chamber photograph is shown in Fig. (5.15). In this example the tracks of the two alpha particles and also that of the associated electron lie in the same plane. In all, twenty-eight out of 10,000 photographs had electron tracks associated with the tracks of the two alpha particles, and could, therefore, be analyzed for "missing" momentum. The component of momentum of the two alpha particles which was perpendicular to the average line of the two particles was measurable through the slight deviation from 180° of the angle between the two tracks. The maximum deviation from 180° was computed to be about 6°. In theory, the component of momentum parallel to the average line of the two alphas could be determined from the difference in the range of the two particles. However, rather large errors were introduced into the range measurements by the fact that at least one of the alphas had to pass through the LiOH target, often obliquely. After investigation of the errors to be expected, the authors concluded that only the perpendicular component of momentum could be measured with sufficient accuracy. It was possible in the 28 cases mentioned above to add the perpendicular component of the momentum of the two alphas to the corresponding momentum component of the electron and to obtain $P_\perp$, the resultant momentum of the three observable particles. This was compared with the possible total momentum of a neutrino of zero rest mass which has an energy given by the maximum energy released in the Li^8 break-up minus the observed energy of the two alphas and the electron.

An examination of the data on these 28 pictures showed that in the majority of cases the observed resultant momentum of the three observable particles was much larger than could be explained on the basis of scattering alone. In many of these cases the momentum could be understood on the neutrino hypothesis. However, the authors concluded

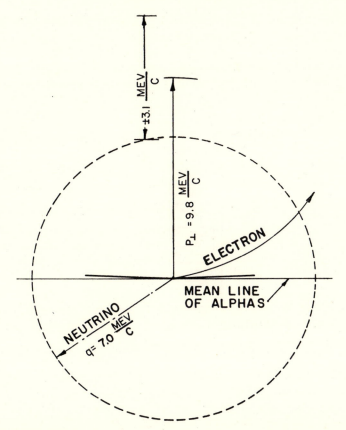

Fig. 5.15. A sketch of a cloud track photograph showing the break up of Li⁸ into two alpha-particles and an electron. The resultand momentum of the three particles is 9.8 ± 3.1 Mev/c, where the uncertainty arises from the possibility of scattering of the alpha particles. A neutrino with the remaining available energy of 7.0 Mev could have this momentum within the probable error. From Christy, Cohen, Fowler, Lauritsen, and Lauritsen.[16]

that the data were insufficient for a meaningful numerical comparison with any assumption concerning the angular distribution of the electron and neutrino.

5.5 Nuclear Recoils from the Beta-Decay of C¹¹

The beta-decay of C¹¹ is given by:

$$C^{11} \to B^{11} + e^{+} + \nu, \qquad T_{1/2} = 20.4 \text{ min}, \tag{5.7}$$

with $E_\beta^{\max} = 0.98$ Mev, $E_0 = 96$ ev, and a spin change $J = \frac{3}{2}$ to $J = \frac{3}{2}$, no change of nuclear parity. According to Table (4.1) this transition is allowed by both Gamow-Teller and Fermi selection rules.

The first attempt in history to carry out a nuclear recoil experiment probably was made by Leipunsky[17] in 1936. This experiment is rather fascinating due to the fact that a satisfactory interpretation of the results is impossible, although the method used must be considered modern, even at the present time. He studied recoils from the beta-decay of C^{11} with the apparatus shown in Fig. (5.16). The C^{11} in the

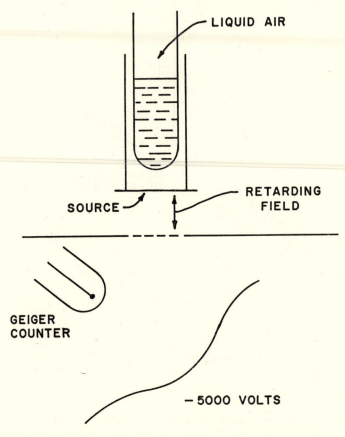

Fig. 5.16. Schematic diagram of the apparatus used by Leipunski[17] to study the recoils from the beta-decay of C^{11}.

form of carbon dioxide or carbon monoxide was condensed on a surface cooled by liquid air. The recoil ions emitted from this source passed through a grid and then were accelerated to a secondary electron-emitting surface maintained at -5000 volts. The secondary electrons ejected from this surface were accelerated to an energy of 5000 ev and then were counted by a Geiger counter with a thin window.

Although the decay of C^{11} to B^{11} by positron emission should inherently produce negative B¹¹ ions, a small yield of positive ions was detected. This is not surprising since several electrons may be ejected from the nucleus as a result of the sudden change of nuclear charge during the beta-decay process. Leipunski's results are shown in Fig. (5.17). The points represent the counting rate observed as a function of the retarding potential between the source and the grounded grid.

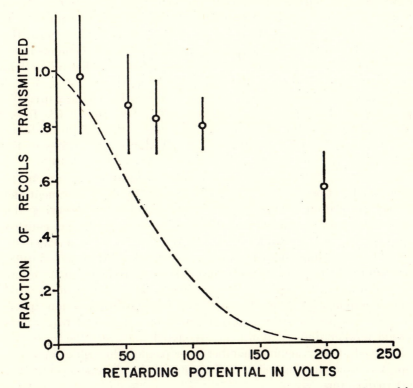

Fig. 5.17. The integral energy spectrum of the recoils from C¹¹ as observed by Leipunski.[17] The dotted curve was computed on the assumption that the momentum of the recoil was equal to that of the electron alone.

Since the maximum energy of the recoil B¹¹ ions should have been about 96 ev, it is obvious that the observed recoil energy spectrum was incorrect. A possible explanation for the apparent presence of ions with energies greater than 96 ev is that the field from the secondary surface at −5000 volts penetrated the grid below the source. This field would have decreased the effect of the applied retarding field, and as a result the apparent energy of the recoils would have been increased. Unfortunately, no information concerning the form of the electron-neutrino angular correlation can be derived from this experiment.

5.6. Angular Correlation in the Beta-Decay of Ne^{19}

Ne^{19} is a positron emitter which decays according to the scheme:

$$Ne^{19} \rightarrow F^{19} + e^{+} + \nu, \qquad T_{1/2} = 18.5 \,\text{sec}, \qquad (5.8)$$

with $E_\beta^{max} = 2.235 \pm 0.005$ Mev, and $E_0 = 205$ ev. The transition occurs between mirror nuclei with spin change $J = \frac{1}{2}$ to $J = \frac{1}{2}$, no change of nuclear parity. According to Table (4.1) this transition is allowed by both Gamow-Teller and Fermi selection rules. Since the results of the He^6 recoil experiments indicate that the Gamow-Teller interaction is T, the interaction in the case of Ne^{19} is either ST or VT. The electron-neutrino angular correlation factor will be $-1 < \lambda < \frac{1}{3}$ for the ST combination, or $+\frac{1}{3} < \lambda < 1$ for the VT mixture. We may consider the beta ray spectrum of Ne^{19} to be simple, since Jones, Phillips, Johnson, and Wilkinson[18] have shown that the probability of a transition to either the 0.112 or 0.200 Mev excited states of F^{19} is less than that to the ground state by a factor of 10^{-2} to 10^{-3}.

The first Ne^{19} electron-neutrino angular correlation experiment to be reported is that of Alford and Hamilton.[19] In this experiment the angular correlation coefficient λ was determined by observing the energy spectrum of nuclear recoils coincident with positrons emitted approximately antiparallel to the recoils. Fig. (5.18) shows a schematic diagram of their apparatus. The positrons were detected by the scintillation counter at the top of the figure. The negative recoil ions which passed downward through the set of triple grids at C were accelerated through a potential difference of 2 kev into an electron multiplier. The Ne^{19} gas produced by a p, n reaction on a target of magnesium fluoride powder diffused down an evacuated line from the cyclotron and arrived at the vacuum chamber after passing through a liquid air trap. A thin conducting foil at "a" separated the chambers containing the two detectors.

A rather novel method involving a combination of transit-time and energy discrimination was used to determine the lower boundary of the effective source volume. A disintegration occurring at a distance L_0 above grid C and producing a positron directed towards the electron counter initiated the following events: At $t = 0$ the positron was detected and a retarding voltage $V(t)$ was applied to the grid at C. This retarding voltage was given by:

$$V = V_0, \, t < t_0, \, (eV_0 = \text{maximum recoil energy})$$

$$V = V_0 \, (t_0/t)^2, \, t > t_0,$$

where $t_0 = L_0(M/2eV_0)^{1/2}$ was the time required for the most energetic recoil to traverse the distance L_0. By means of this variable retarding

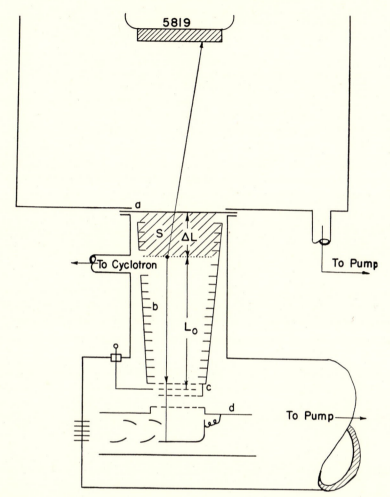

Fig. 5.18. A schematic diagram of the apparatus used by Alford and Hamilton[19] in their study of the electron-neutrino angular correlation in the decay of Ne¹⁹. The effective source region is denoted by S.

voltage, recoils originating below the lower boundary of the source were prevented from passing grid C, whereas recoils from above this boundary could pass through. For a given t_0 the sensitive volume was thus limited to the region $L_0 < L < L_0 + \Delta L$. The momentum spectrum of the recoil nuclei was computed from the distribution in the transit times of the ions which passed through the retarding grid.

The shape of the recoil spectrum observed in this experiment corresponded to an angular correlation factor of $\lambda = -0.8 \pm 0.4$. As mentioned earlier, an angular correlation factor within the range $-1 < \lambda < \frac{1}{3}$ indicates that the Fermi part of the beta interaction is scalar. Although the value $\lambda = -0.8 \pm 0.4$ clearly indicated the presence of the scalar interaction, this value was in strong disagreement with the value of $\lambda = -0.06 \pm 0.02$ expected from the comparative half-life ft of the Ne[19] beta-decay.

Alford and Hamilton[46] have carried out a second Ne[19] experiment in which the angular correlation was obtained by observing the spectrum of positrons coincident with nuclear recoils emitted approximately antiparallel to the positrons. The electron-spectrum measurements depended on the time of flight of the recoils only for localization of the source volume in which the recoils were produced. This localization of the source volume limited the range in the values of the electron-neutrino angle.

The apparatus used in this second experiment was essentially the same as that shown in Fig. (5.18). The most important change was the addition of a 20-channel pulse-height analyzer for the purpose of recording the energy spectrum of the positrons.

The energy spectrum of disintegration positrons coincident with the recoils obtained in the second experiment of Alford and Hamilton is shown in Fig. (5.19). The best fit between the experimental data and the theoretical curves was obtained for an angular correlation $\lambda = -0.15 \pm 0.2$. The authors conclude that the discrepancy between their two values of λ was caused by small inaccuracies in timing the recoil delays, and also in the shape of the pulse applied to the grid

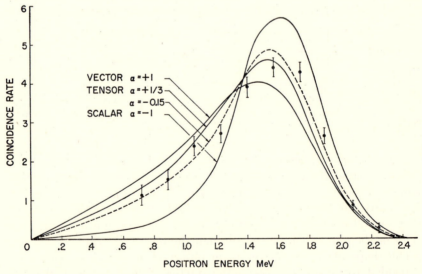

Fig. 5.19. The spectrum of the disintegration electrons from the beta-decay of Ne[19] coincident with recoils at angles near 180°. All curves have been normalized to the same area. From Alford and Hamilton.[46]

near the ion counter. These inaccuracies were greatly reduced in the second experiment. An additional source of error was also eliminated in the second experiment. Since the region accessible to decays giving rise to true coincidences between positrons and neutral recoils was not limited by the timing system, it was necessary to subtract these spurious coincidences from the true coincidences. The isolation of these spurious coincidences was effected by placing a retarding potential of -225 volts on grid C in front of the ion counter. This retarding potential was sufficient to stop the charged recoil ions. The excess coincidences above the chance background were measured and this excess was then subtracted from the measured spectrum.

Maxson, Allen, and Jentschke[20] have measured the electron-neutrino angular correlation in the beta-decay of Ne¹⁹ by observing the energy spectrum of the recoil ions emitted in coincidence with the positrons. The positrons were detected only if they were emitted in an angular range of approximately 45° to 225° from the direction of the beam of recoil ions. A spherical electrostatic spectrometer was used to measure the energy of the recoils. In addition to the measurement of the energy, an estimate of the velocity of the recoil ions was obtained from the measurement of the time of flight of these ions. As a result of these two measurements the recoil ions were identified by their mass-to-charge ratio. An angular correlation of $\lambda = -0.21 \pm 0.08$ was derived from the experimental data.

A schematic diagram of the apparatus used by Maxson et al. is shown in Fig. (5.20). The recoil ions which passed through the electrostatic energy analyzer $G - G$ originated in a roughly hemispherical source volume at E. The source volume was partially surrounded by the plastic scintillator at B. The actual source volume was defined on the side nearest the electron detector by a hemispherical bubble of plastic foil with an aluminized inner surface. The opposite side of the source was defined by a flat sheet of aluminum foil supported on a movable gate of the same type as that used by Allen and Jentschke[14] in their He⁶ experiment. Delayed coincidences between the positrons and the recoil ions were recorded with the gate open and closed. The difference between these two sets of data represented the coincidence rate due to disintegrations within the source volume. In order to prevent the detection of positrons which could not produce true coincidences, the region between the source volume and the scintillator was sealed off from the main vacuum chamber and evacuated with a separate pump. Differential pumping was used to reduce the pressure of the Ne¹⁹ gas in the ion counter.

The Ne¹⁹ was produced by the reaction $F^{19}(p, n)Ne^{19}$, with SF₆ gas serving as both the target and carrier. The carrier gas was removed from the Ne¹⁹ by condensation in a liquid nitrogen trap. A second trap containing chips of calcium heated to 400°C was used to remove chemically active impurities which passed through the cold trap. The Ne[19] gas entered the chamber through tube A of Fig. (5.20) and diffused throughout the vacuum system.

The observed recoil energy spectrum and also the expected distributions calculated for various electron-neutrino angular correlations are shown in Fig. (5.21). A least squares fit of the theoretical curves to the experimental data yielded $\lambda = -0.21 \pm 0.08$, where the indicated error is statistical. This value of the angular correlation clearly indicates that the Fermi part of the beta interaction is scalar, at least in the case of the Ne[19] decay.

The results of a search for positive recoil ions are shown in Fig. (5.22). The time-of-flight spectra of both positive and negative F[19] ions were observed with the gate open. The mean energy of the ions was 170 ev. A comparison of the numbers of counts within the two peaks re-

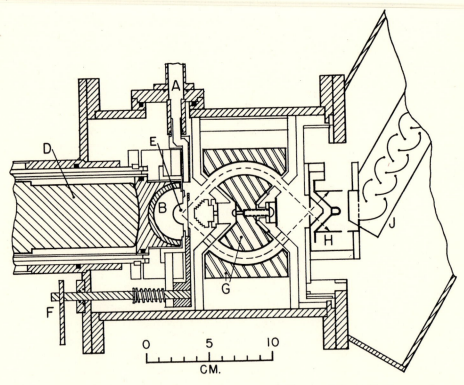

Fig. 5.20. The recoil chamber used by Maxson, Allen, and Jentschke[20] to study the decay of Ne[19]. The positron detector was a scintillation counter with a hemispherical plastic scintillator at B. The ion detector was an electron multiplier at H and J. The negative recoil ions were accelerated through a potential difference of 2000 volts between the conical grids at H.

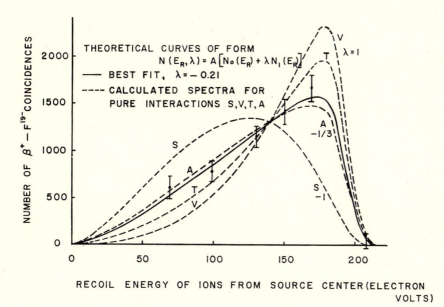

Fig. 5.21. Energy spectrum of the F¹⁹⁻ recoil nuclei from the Ne¹⁹ decay. From Maxson, Allen, and Jentschke.[20]

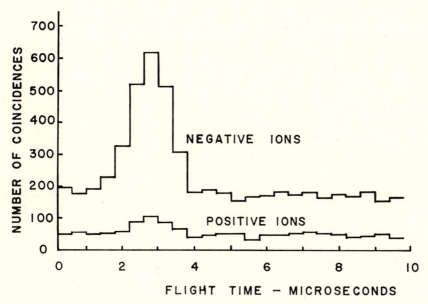

Fig. 5.22. Time-of-flight distributions of positive and negative F¹⁹ recoil ions of 170 ev energy. From Maxson, Allen, and Jentschke.[20]

vealed that the true coincidence rate due to positive recoils was about ten percent of that due to negative recoils. For the production of singly charged positive recoil ions, two electrons must have been ejected from the F^- recoils. The primary cause of this additional ionization is the sudden change of nuclear charge during the beta-decay process. A vacancy produced in the K shell by this "shake up" process may be filled by an electron from one of the L subshells with the concomitant emission of another L electron. The distribution of charge states among the recoils resulting from the decay of A^{37} by orbital electron capture has been measured by Kofoed-Hansen[21] and also by Snell and Pleasonton.[22] Both groups observed that the most probable charge state was three, and that in a few cases up to seven electrons had been ejected. Snell[23] also studied the charge spectrum of the recoils resulting from the negative beta-decay of $Kr^{85} \rightarrow Rb^{85}$, and found that 4.5 percent of the ions carried a charge of three. This corresponded to the removal of two electrons from each Rb^{85} atom. It is apparent from these observations that multiply-charged recoil ions will appear in both positive and negative beta-decay. Recoil experiments should be carried out in such a manner that the interpretation of the results is not affected by the possible presence of either neutral or multiply-charged recoils.

Good and Lauer[47] have measured the electron-neutrino angular correlation in the beta-decay of Ne^{19} with an experimental arrangement which differs in several respects from the arrangements used in other Ne^{19} experiments. In this latest experiment the distribution in kinetic energy of the recoil ions associated with positrons of known energy was measured, and a value of the angular correlation factor λ was deduced from the shape of this energy spectrum. A value of $\lambda = +0.14 \pm 0.13$ was obtained in this experiment.

Because of the geometry of the experiment a wide range of angles between the directions of motion of the positron and of the recoil ion were accepted. The positron energy and the recoil energy were sufficient to completely determine each decay, and therefore permitted the calculation of the positron-neutrino angle for each event. The kinetic energy of each positron was measured with a scintillation counter and the associated recoil ion kinetic energy was determined from a measurement of the time required for the ion to traverse the path from the source to the ion counter.

Fig. (5.23) shows the recoil chamber, the scintillation counter, and the recoil-ion counter used in the experiment of Good and Lauer. A perforated plate separated the "good" source volume near the plastic scintillator from the rest of the vacuum system. This plate offered impedance to the flow of Ne^{19} gas from the good source volume into the drift space. As a result of this restriction to the flow of gas, the

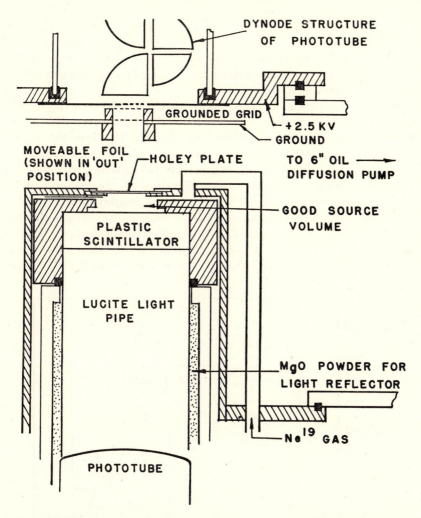

Fig. 5.23. The vacuum chamber, recoil ion counter and scintillation counter used by Good and Lauer[47] in their measurement of the electron-neutrino angular correlation in the beta-decay of Ne¹⁹.

density of the active gas in the drift space was about 1/100 of that in the effective source volume. A movable foil was used to subtract from the true effect the background of delayed coincidences due to decays in the drift space. Since this foil was transparent to the positrons and opaque to the recoil ions, the difference between the number of coincidences observed with this gate open and closed represented the number of events occurring in the good source volume.

The time of flight of the recoil ions was measured by recording coincidences between the prompt pulses caused by the detection of positrons in the scintillator and the delayed pulses due to the arrival of the associated recoil ions at the ion counter. The prompt pulses started the horizontal sweep of an oscilloscope, and the delayed recoil pulses were displayed as vertical deflections. A 35-mm strip camera photographed the pairs of pulses.

The experiment was run for approximately five 16-hour days, alternating between "foil open" and "foil closed" about every 25 minutes. The data taken in this run are shown in Fig. (5.24), where the number of positron-recoil ion coincidences have been plotted as a function of cos $\theta_{e\nu}$. As mentioned earlier, $\theta_{e\nu}$ cannot be measured directly, but must be obtained from the measured values of the energies of the positron and associated recoil ion. Cos $\theta_{e\nu}$ is roughly proportional to the recoil energy. A least-squares fit of the theoretical curve to the data of Fig. (52.4) gave $\lambda = +0.14 \pm 0.13$, which again suggests that the Fermi component of the beta-decay interaction is scalar. This result is considerably more positive than the value $\lambda = -0.15 \pm 0.2$ obtained

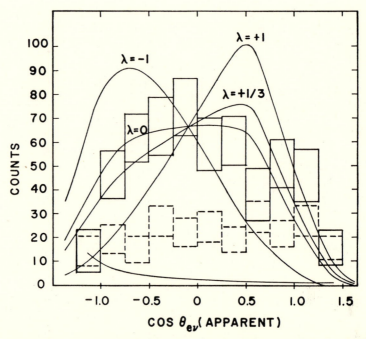

Fig. 5.24. The distribution in cos $\theta_{e\nu}$ of the positron-recoil ion coincidences from the decay of Ne[19]. Cos $\theta_{e\nu}$ is roughly proportional to the kinetic energy of the recoil ions. Solid rectangles give the measured data, with random background and foil-closed background subtracted. The dotted rectangles are foil-closed background with foil-closed randoms subtracted, and the solid curve R is the foil-open random background. The solid curves are theoretical with instrumental resolution folded in. The best fit between the experimental data and the theoretical curves was obtained with $\lambda = +0.14 \pm 0.13$. From Good and Lauer.[47]

by Alford and Hamilton, or the value $\lambda = -0.21 \pm 0.08$ obtained by Maxson et al. A possible explanation of this discrepancy is that the ion detector used by Lauer was sensitive to neutral recoil atoms. Since the detection efficiency would be expected to increase with increasing kinetic energy of the neutral atom, this effect would bias the data toward more positive values of λ. Apparently the data of this experiment were not corrected for this effect.

5.7. Angular Correlation in the Beta-Decay of P^{32}

The beta-decay of P^{32} is given by:

$$P^{32} \to S^{32} + e^- + \nu, \qquad T_{1/2} = 14 \text{ d.}, \tag{5.9}$$

with $E_\beta^{\max} = 1.72$ Mev, $E_0 = 78$ ev. According to shell theory the nuclear configuration of the P^{32} ground state is $s_{1/2}d_{3/2}$. Enge[24] has assigned a spin of $J = 1+$ to this state, based on the assumption that the P$^{31}(d, p)$P^{32} yield is proportional to $2J + 1$. Since the ground state of the even-even nucleus S^{32} probably has $J = 0+$, the P$^{32} \to S^{32}$ transition has $\Delta J = 1$, no change of nuclear parity, and should be allowed. However, the fact that the ft value for P^{32} is about 10^4 times that for a typical allowed decay suggests that the transition is actually "l" forbidden. Apparently the $d_{3/2}$ neutron in P^{32} changes to an $s_{1/2}$ proton in the S^{32} nucleus, with a change $\Delta l = 2$.

Sherwin[25] has studied the decay of P^{32} in a series of three experiments. In these experiments the momentum of the electron, the momentum of the associated recoil ion, and also the electron-recoil angle were measured. Since the simultaneous measurement of the momentum of the electron and the electron-recoil ion angle uniquely defined the momentum of the nuclear recoil, a comparison could be made between the measured and predicted recoil momentum. Most of the sources exhibited broad momentum peaks instead of the sharp peaks which had been predicted. However, a few sources did show sharply defined momentum peaks at several values of the electron-recoil ion angle. The data obtained from these good sources were converted to a distribution of recoil momenta as a function of the angle between the electron and neutrino. The experimental points were fitted adequately by an angular correlation of the approximate form $\lambda = 1 + \cos \theta_{e\nu}$.

These experiments were carried out with great care, and indicate the difficulties which arise when the source is a thin film. The sources were prepared by successively coating a glass film (300 to 500 $\mu g/cm^2$) with a barely visible film of aluminum, 100 to 1000 atomic layers of lithium fluoride, and finally with less than one atomic layer of P^{32}. The entire fabrication of the source was carried out in part of the vacuum chamber used for the recoil experiments. In order to prevent the absorption of gas on the thin film of P^{32}, the source was held at a temperature of about 100°C. The LiF substrate provided a high work-function surface which did not neutralize the positive S^{32} recoil ions. No recoil ions were observed when the P^{32} was deposited directly on the aluminum. The conducting sublayer of aluminum was required to keep the surface of the source on the LiF at a constant potential.

The apparatus used by Sherwin is shown in Fig. (5.25). The thin layer of P^{32} at S faced the electron multiplier used as the ion counter. The electrons passed through the source backing and were analyzed by four 90° magnetic spectrometers at various angles with respect to the direction of the recoils. The spectrometers were all operated at the same magnet current and selected the same momentum interval. A time-of-flight method was used to measure the distribution in the velocity of the recoil ions.

Most of the P^{32} sources gave no useful information about the neutrino. Apparently the recoil ions were strongly scattered upon leaving the

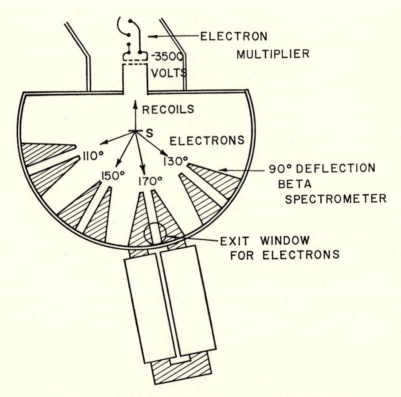

Fig. 5.25. The experimental apparatus used by Sherwin[25] to study the electron-neutrino angular correlation in the decay of P^{32}. The four beta spectrometers were all in the same vacuum and had a common focus at the source S.

surface of the source. As a result of this scattering, the momentum distribution of the recoils had a broad peak instead of the sharply defined peak expected as a resultant of the vector sum of the selected momenta of the electron-neutrino pair.

A few of the sources apparently were very thin and showed sharp momentum peaks. The momentum distribution of the recoil ions from a typical good source is shown in Fig. (5.26). Definite peaks were observed for recoils with momenta of 6950 and 6150 $H\rho$. However, the expected sharp peak was not observed for the recoils with momenta of 4960 $H\rho$. It is evident from these curves that it was very difficult to obtain a reliable estimate of the actual number of recoil ions emitted within a given angular range from the direction of motion of the elec-

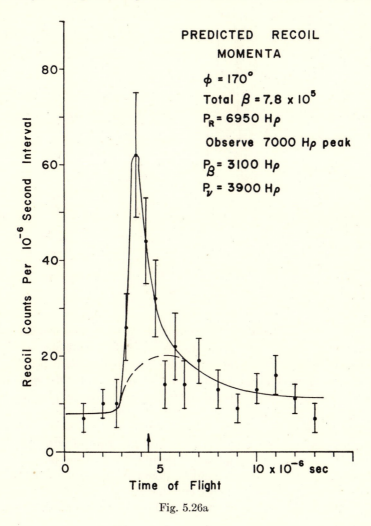

PREDICTED RECOIL MOMENTA

$\phi = 170°$

Total $\beta = 7.8 \times 10^5$

$P_R = 6950\ H\rho$

Observe 7000 $H\rho$ peak

$P_\beta = 3100\ H\rho$

$P_\nu = 3900\ H\rho$

Fig. 5.26a

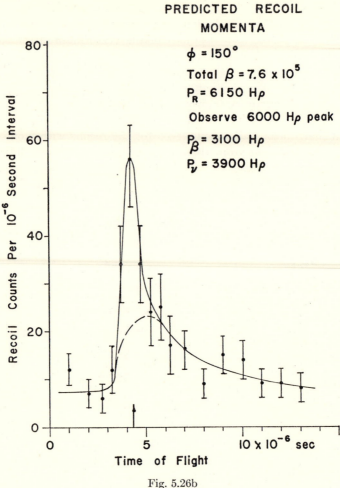

PREDICTED RECOIL MOMENTA

$\phi = 150°$

Total $\beta = 7.6 \times 10^5$

$P_R = 6150$ Hρ

Observe 6000 Hρ peak

$P_\beta = 3100$ Hρ

$P_\nu = 3900$ Hρ

Fig. 5.26b

trons. The data showed that the fraction of the recoil ions which escaped the surface with little scattering was very small. About one percent of the expected number were actually observed in the sharp peaks.

An unambiguous interpretation of the results of this experiment probably is impossible at present. Unfortunately, the complete explanation of the "l" forbidden beta-decay of P^{32} has not been worked out. For example, Porter et al.[26] may have found evidence for small deviations from the allowed shape in the beta-spectrum of P^{32}. A possible explanation is that through some peculiarity of the nuclear wave functions, the $|\langle \beta\sigma \rangle|^2$ which should be dominant in an allowed $\Delta J = 1$ transition is suppressed to the extent that the usually neglected higher order matrix elements[49] become important. In general these matrix

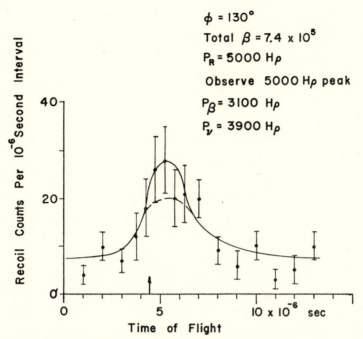

PREDICTED   RECOIL
MOMENTA

$\phi = 130°$

Total $\beta = 7.4 \times 10^5$

$P_R = 5000$ Hρ

Observe 5000 Hρ peak

$P_\beta = 3100$ Hρ

$P_\nu = 3900$ Hρ

Fig. 5.26c. Recoil ions from a good P³² source. Superimposed on the background due to chance events and to scattered ions, are sharp groups of recoil ions which occur at times which are in qualitative agreement with the predicted values. The short arrows on the time-of-flight axis show where the peaks are expected. From Sherwin.[25]

elements depend upon the energy of the electron, and consequently will affect the shape of the beta-spectrum. B. C. Carlson in *Iowa State College Report ISC-758* reports some preliminary estimates for the case of C^{14}, in which the interference between $\langle \beta \sigma \rangle$ and the retarded matrix elements, particularly $\langle \beta \alpha \times \mathbf{r} \rangle$ makes remarkably large contributions compared to $|\langle \beta \sigma \rangle|^2$, and introduces a weak-energy dependence in the shape factor of the beta-spectrum. This type of interference may account for the small deviation from the allowed shape in the case of P^{32}. The presence of these interference terms may affect the electron-neutrino angular correlation to a considerable degree. This problem is now under investigation, but results are not yet available.

5.8. Nuclear Recoils from the Beta-Decay of Cl^{38}

The beta-decay of Cl^{38} is given by:

$$Cl^{38} \rightarrow A^{38} + e^- + \nu, \qquad T_{1/2} = 38 \text{ min}, \qquad (5.10)$$

with $E_\beta^{max} = 5$ Mev and $E_0 = 424$ ev. The transition to the ground state of A^{38} has been classified as a unique, first forbidden type, and consequently the electron-neutrino angular correlation is given by equation (4.8).

Crane and Halpern[28] used the cloud chamber method to study the nuclear recoils from the decay of Cl^{38}. The radioactive isotope was introduced into the cloud chamber as ethylene dichloride vapor. When the chamber was expanded immediately after the beta disintegration, the very short track of the recoil nucleus appeared only as a point at the beginning of the electron track. The nucleus did, however, produce a small number of ion pairs, and this number was some function of the kinetic energy of the nucleus. In order to view this group of ion pairs, the electric clearing field was removed and the expansion of the chamber was delayed for about $\frac{1}{2}$ second after the beta disintegration. By this time the cluster of ions had spread by diffusion into a spherical region several millimeters in diameter, and the number of ion pairs could be determined by counting the droplets. A measurement of the curvature of the electron track in a magnetic field gave the momentum of the electron associated with the recoil ion.

Unfortunately, the decay of Cl^{38} is complex with three beta-spectra of 5 Mev (53%), 2.8 Mev (16%), and 1.1 Mev (31%). Apparently the authors were unaware of the presence of the 2.8 Mev component and eliminated only the 1.1 Mev component in their measurements.

Fig. (5.27) shows the distribution in the number of droplets produced by the recoils from the decay of Cl^{38}. If we assume that two droplets were produced for each 30 ev loss of energy, the maximum number of droplets observed should have been about 28. Apparently in some disintegrations, about twice the expected number of droplets were produced. The authors suggest that these additional droplets were formed on the neutral atoms or molecules resulting from the dissociation of the original molecule containing the Cl^{38}. Since the amount of dissociation may have been some function of the energy of the recoil, the number of droplets probably was not proportional to the energy of the recoil. According to the unique first forbidden character of the $Cl^{38} \rightarrow A^{38}$ transition, the recoil energy distribution should be peaked at the high-energy end. In contrast to this peaked distribution, the data of Fig. (5.27) indicate an almost constant distribution of the energy of the recoils. Apparently the true distribution was distorted by the effects of molecular dissociation as mentioned earlier. Additional distortion may have resulted from the relatively high probability of the transfer of momentum from the recoiling nucleus to the fragments of the original molecule. In principle, this type of recoil experiment utilizing a radioactive atom in a molecular host could be designed to give information regarding the breakup of the molecule during the disintegration.

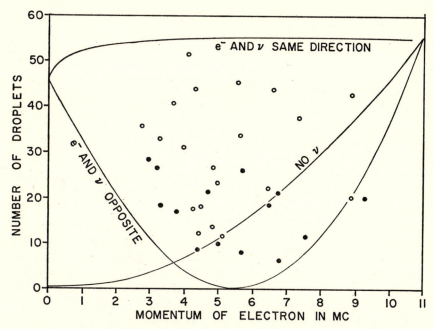

Fig. 5.27. The results obtained by Crane and Halpern[28] in their cloud chamber study of the decay of Cl³⁸. Each dot or circle represents a droplet cluster.

5.9. Angular Correlation in the Beta-Decay of A^{35}

A^{35} is a positron emitter which decays according to the scheme:

$$A^{35} \rightarrow Cl^{35} + e^+ + \nu, \; T_{1/2} = 1.83 \text{ sec}, \tag{5.11}$$

with $E_\beta^{\max} = 4.96 \pm 0.04$ Mev and $E_0 = 455$ ev. The spin assignment for the ground states of these mirror nuclei is $J_i = +\frac{3}{2}$ and $J_f = +\frac{3}{2}$. Kistner et al.[59] have made a careful investigation of the positron-decay of A^{35} and have discovered that the beta-spectrum is complex. Analysis of the complex beta-spectrum showed that approximately seven percent of the decays go to two excited states in Cl^{35}. According to their measurements the ft value of the ground-state, mirror transition is 6200 $\pm$ 400 seconds. This ft value clearly indicates that this transition is allowed. A^{35} is particularly well suited for identifying the Fermi invariant in the beta-decay interaction, since the Gamow-Teller nuclear matrix elements are small compared to the Fermi matrix elements. Since the Fermi nuclear matrix element $|\langle 1 \rangle|^2$ is very nearly equal to unity for the mirror type ground state transition, the Gamow-Teller matrix element $|\langle \sigma \rangle|^2$ can be computed from ft values according to the procedure described in Section 5.12. An evaluation by this method yields $|\langle \sigma \rangle|^2 \cong 0.05$. In view of the small value of $|\langle \sigma \rangle|^2$, the angular correlation factor (equation 4.5) should be nearly $+1$ if the Fermi part of the beta interaction is vector, or nearly -1 if the Fermi part is scalar. Since the shapes of the energy spectra of recoil nuclei differ strongly for these extreme values of the angular correlation, an electron-neutrino recoil experiment should be able to clearly distinguish between these two possible interactions.

Maxson, Herrmannsfeldt, Stahelin, and Allen have measured the electron-neutrino angular correlation in the beta-decay of A^{35} using the apparatus shown in Fig. (5.20). This same apparatus had been used previously to study the electron-neutrino angular correlation in the decay of Ne^{19}. The average value of the angular correlation obtained from four groups of A^{35} experiments was $\lambda = +0.85 \pm 0.12$.

The A^{35} was produced by the reaction $S^{32}(\alpha, n)A^{35}$, with SF_6 gas serving as both the target and carrier. As in the Ne^{19} experiment, the carrier gas was removed from the active gas by condensation in a liquid nitrogen trap. A trap containing hot calcium chips was used to remove chemically active impurities which passed through to a cold trap. The purity of the radioactive gas was checked by life-time measurements. No significant amounts of positron emitting impurities were found with the exception of a small amount of an activity with a half-life of approximately 18 seconds. This impurity may have been Ne^{19} produced by the reaction $O^{16}(\alpha, n)Ne^{19}$. The target material probably was water vapor which had slowly collected in the gas target system.

The value of the angular correlation observed in this first A^{35} experiment clearly indicated that the vector invariant is the dominant part of the Fermi interaction in this particular beta-decay. Since the results of the He^6 experiment together with those of the neutron or Ne^{19} experiments apparently indicated that the beta-decay interaction was S, T rather than V, T, a decision was made to repeat the A^{35} experiment with a new apparatus.

The second A^{35} experiment was performed by Herrmannsfeldt, Stahelin, and Allen.[58] In this experiment the energy distribution of the recoil ions was measured directly without a simultaneous measurement of either the direction of motion or the energy of the associated electrons. As shown in Fig. (4.1), the energy distribution measured in this manner depends rather strongly upon the form of the electron-neutrino angular correlation. Since this method did not require the counting of recoil ion-electron coincidences, there was no intensity limitation due to a background of chance coincidence counts. In addition, this method eliminates the uncertainties inherent in measurements of the direction of motion of the electron. The results of this experiment were essentially the same as those of the first A^{35} experiment.

A schematic diagram of the apparatus used in the second investigation of A^{35} is shown in Fig. (5.28). The energy analysis was performed by

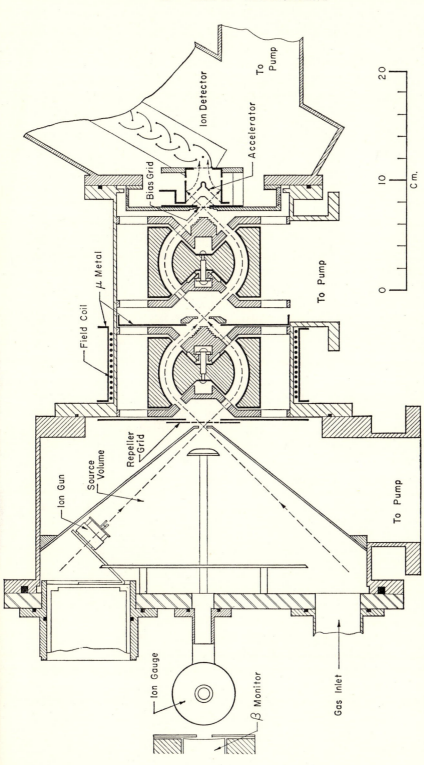

Fig. 5.28. A schematic diagram of the apparatus used by Herrmannsfeldt et al.[58] to measure the electron–neutrino angular correlation in the beta-decay of A³⁵. The effective source volume is inside the cone at the left of the diagram. The energy measurements were made by two electrostatic spectrometers in series. Differential pumping was used to reduce the pressure of the radioactive gas in the recoil ion counter. A retarding potential could be applied to a grid in front of the exit aperture of the source volume in order to separate the background from the true effect.

two electrostatic spectrometers in series. The effective source volume for the first spectrometer was a volume inside the cone at the left of the diagram. The outer part of the source volume was defined by the metal wall of the cone, and the inner part by the acceptance angles of the first spectrometer. A pressure differential between the source volume and the spectrometers was maintained by a series of apertures and vacuum pumps. For most of the runs the total pressure in the source volume was about 3×10^{-5} mm of Hg.

Despite the use of differential pumping, the background counting rate was perhaps ten times the maximum true rate. In order to measure this background, a set of grids was placed in front of the exit aperture of the source volume. A retarding potential could be placed on the inner grid to shut off the beam of ions coming from the source volume. The counting rate observed with this retarding potential applied represented the background rate. During the experimental observations a switching system automatically switched the retarding potential on and off and also switched the output of the ion counter alternately between two recording circuits. The difference between these two sets of accumulated counts represented the true effect for a given setting of the spectrometers. The source activity was monitored by a Geiger-Müller counter, which also was switched alternatively between two recording circuits as the retarding potential was turned on and off.

Preliminary tests showed that low-energy electrons passed through the spectrometers since the polarity of the deflecting and accelerating voltages was selected for the detection of negative ions. A single layer coil outside the first spectrometer provided a weak magnetic field which effectively prevented the transmission of electrons. Tests with an ion source showed that the transmission of the negative ions would not be seriously affected.

The target and gas purification system were identical with those used in the first A^{35} experiment. However, during the later runs of the second experiment the SF_6 gas was circulated over a drying agent to remove water vapor.

The recoil energy spectrum observed in the second A^{35} experiment and also the distributions calculated for various assumed electron-neutrino angular correlations are shown in Fig. (5.29). The theoretical curves have been corrected for the finite resolution of the spectrometers and for the contribution of the seven percent branch of the A^{35} decay. The second correction could be made only approximately, since the seven percent branch actually consists of a two percent and a five percent branch, each going through an excited state to the ground state of Cl^{35}. The comparative half-lives indicate that both transitions are allowed. Since the two excited states have about the same energies, the corrections to the recoil spectra were based on a single transition to an excited state midway between the two actual states. Since the contribution to the primary recoil energy spectrum is smeared out due to the presence of the momentum of the gamma ray which follows a transition to an excited state, the correction does not depend strongly on the type of beta interaction. Shell theory suggests that the spins of the two excited states are $J = \frac{1}{2}$ and $J = \frac{5}{2}$. According to this assignment the spin change for either branch would be $|\Delta J| = 1$, which would rule out a transition with Fermi matrix elements. The corrections were

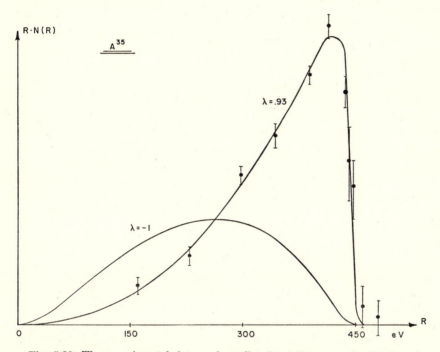

Fig. 5.29. The experimental data and predicted recoil energy distributions for A³⁵. The theoretical curves have been corrected for the finite resolution of the spectrometer, and also for the seven percent branch of the A³⁵ decay. The experimental points correspond to an angular correlation factor of $\lambda = +0.93 \pm 0.14$. All curves were normalized to the same area before being corrected for the finite resolution of the spectrometer. From Herrmannsfeldt et al.[58]

made with the assumption of the axial vector interaction, but the use of the tensor interaction would not appreciably change the shape of the curves. All curves in Fig. (5.29) were normalized to the same area before being corrected for the resolution of the spectrometer.

The experimental points shown in Fig. (5.29) were fitted to the computed curves by a least squares fit. The best fit was obtained for $\lambda = +0.93 \pm 0.14$, where the quoted error is such that there is a probability of about $\frac{2}{3}$ that a measured value falls within the limits of error.

A chi-squared test showed that the data were statistically significant.

It is evident from Fig. (5.29) that the angular correlation again strongly suggests that the Fermi part of the beta-decay interaction is vector.

5.10. Nuclear Recoils from the Beta-Decays of Kr^{88} and Kr^{89}

Kr^{88} decays as follows:

$$Kr^{88} \xrightarrow[2.7hr]{e^-} Rb^{88} \xrightarrow[17.8min]{e^-} Sr^{88}. \qquad (5.12)$$

Thulin[30] has measured the shape of the complex beta-spectrum of Kr^{88} and has found three components of energy: 2.8, 0.9, and 0.52 Mev, with intensities of 20, 12 and 68 percent, respectively. Because of the low intensity, the existence of the 0.9 Mev component was not considered as well established. He has also investigated the shape of the Rb^{88} beta-spectrum and has concluded that the ground state of Rb^{88} has a spin of $J = 2-$. Since the Kr^{88} nucleus is even-even, the spin probably is $0+$. From these considerations the beta component due to a transition from Kr^{88} to the ground state of Rb^{88} should have the same unique first forbidden character as the beta-decay of Cl^{38}.

The decay of Kr^{89} and its daughter substances can be summarized as follows:

$$Kr^{89} \xrightarrow[3.18min]{e^-} Rb^{89} \xrightarrow[15.4min]{e^-} Sr^{89} \xrightarrow[55d.]{e^-} Y^{89}. \qquad (5.13)$$

Kofoed-Hansen and Nielson[31] have found the decay of Kr^{89} to be complex, with two beta-spectra having end points of 3.9 and ≈ 2 Mev, respectively. The lower limit for the branching ratio of the 2 Mev group was about 35 percent.

Jacobsen and Kofoed-Hansen,[32] and later Kofoed-Hansen and Kristensen[33] have studied the nuclear recoils from the decay of Kr^{88} and also Kr^{89} using a method entirely different from the experimental procedures described in this chapter. In the Kofoed-Hansen experiments the maximum energy and also the mean energy of the recoils were determined using electrostatic retarding potential methods. Since in each case the Kr isotope decayed to a radioactive daughter, the recoils could be collected on a foil which was assayed at a later time.

The apparatus used for the measurement of the maximum recoil energy is shown in Fig. (5.30). A box-shaped structure *EFG* with a grid-covered opening on one side was supported by two insulating flanges *C*. The flanges carried two metal plates *D* and the two collectors *A* and *B*. In order to carry out a measurement of the recoil energy the radioactive gas was introduced into the chamber, and a retarding potential applied between the box and the collector plates. The gas was allowed to remain in the system for several minutes and then was removed, and the activity collected on plates A and B was assayed. This process was repeated for a suitable range of retarding potentials. A plot of the ratio of the activities N_A/N_B of the plates A and B as a function of the retarding voltage gave a curve which decreased with

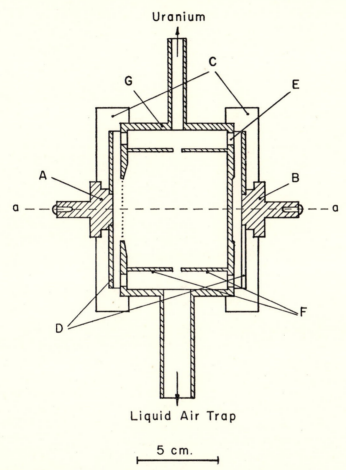

Fig. 5.30. The apparatus used by Kofoed-Hansen and Kristensen[32] to measure the maximum energy of the recoils from the beta-decay of Kr⁸⁹. The recoil nuclei were collected on plates A and B.

increasing voltage and reached a constant value when the stopping potential exceeded the maximum energy of the recoil ions. In principle, this type of measurement should give the maximum energy E_0 of the particles, but corrections must be applied to take care of the transparency of the grid.

More definite information about the energy distribution of the recoils was obtained from a determination of the average value of the energy. The apparatus used is shown in Fig. (5.31). This consisted of a plane-parallel condenser placed in a vacuum tight box. One of the plates was insulated from the box and placed at a potential of $+V$ volts with respect to the box. The radioactive gas was introduced into the chamber for a suitable time and then the activities of the collector plates were measured. When $Ve > E_0$, the ratio of the number of recoils collected on the positive

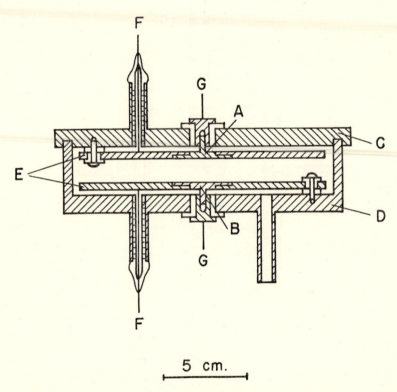

5 cm.

Fig. 5.31. The apparatus used by Kofoed-Hansen and Kristensen[32] to measure the average energy of the recoil nuclei from the beta-decay of Kr^{89}. When a retarding potential of $+V$ volts was applied to the upper plate, the average energy of the recoils could be computed from the ratio of the activities collected on the two plates.

plate $N+$ to the number collected on the negative plate $N-$ gave a direct measurement of the energy E_R divided by the charge Z_R of the recoils. This connection is given by:

$$\langle E_R/Z_R \rangle = 6V(N_+/N_-)/(1 + N_+/N_-),\qquad(5.14)$$

where $\langle\ \rangle$ denotes the average value. As pointed out by the authors, this method will not yield the correct $\langle E_R \rangle$ unless the distribution of charge carried by the recoils is known. For example, Snell[23] has found that the nuclear recoils resulting from the decay $Kr^{85} \xrightarrow[9.4y]{e^-} Rb^{85}$ on the average carried a charge of 1.5 electronic units. This suggests that the $\langle E_R/Z_R \rangle$ measured in the Kofoed-Hansen experiments should be multiplied by approximately 1.5 to obtain a correct value of $\langle E_R \rangle$.

In the Kr^{88} experiment the measured average recoil energy $\langle E_R/Z_R \rangle$ was 29 ± 1 ev, and the measured value of the maximum recoil energy E_0 was 51.5 ± 2 ev. The observed value of E_0 was considerably lower than the expected value of 67 ev for a maximum electron energy of 2.8 Mev. Since the measured value of $\langle E_R/Z_R \rangle$ included the contribution from the relatively abundant low-energy components of the Kr^{88} beta-spectrum, the average recoil energy from the 2.8 Mev beta-spectrum alone would have been considerably higher than 29 ev. In addition, the average energy would have been further increased if the correction for the average charge had been applied. Since the average recoil energy apparently is greater than half the maximum value, the spectrum probably is peaked at the high-energy end. This shape is of the form expected for a unique, first forbidden transition with either the T or A interaction.

The results of the Kr^{89} experiment were similar to Kr^{88} data. The maximum recoil energy of 115 ± 5 ev corresponded to a maximum electron energy of 3.9 ± 0.1 Mev, which is in good agreement with the directly measured value of 4.0 Mev. The measured value of $\langle E_R/Z_R \rangle$ was 58 ± 2 ev. Kofoed-Hansen and Kristensen have estimated the contribution to $\langle E_R/Z_R \rangle$ from the 2 Mev component of the Kr^{89} spectrum. This correction increased the average recoil energy to about 72 ev. This represents a lower limit to $\langle E_R \rangle$ since $\langle Z_R \rangle \cong 1.5$ rather than 1.0. The authors concluded that the recoil spectrum resulting from the transition of Kr^{89} to the ground state of Rb^{89} was of the shape expected for an angular correlation of the same form as equation (4.8).

5.11. Angular Correlation in the Beta-Decay of Y^{90}

The decay of Y^{90} is given by:

$$Y^{90} \rightarrow Zr^{90} + e^- + \nu, \qquad T_{1/2} = 62 \text{ hrs}, \qquad (5.15)$$

with $E_\beta^{max} = 2.26$ Mev, and $E_0 = 44$ ev. The Y^{90} beta-spectrum departs noticeably from the allowed shape and corresponds to a unique forbidden transition with an electron-neutrino angular correlation given by equation (4.8).

Sherwin[34] has studied the momentum spectra of the recoils from monolayer sources of Y^{90}. The apparatus was essentially the same as that used in the P^{32} experiments described in Section 5.7, with the exception that the direction of the electrons was measured without a simultaneous measurement of the momentum. The thin layer of Y^{90} was deposited by vacuum evaporation upon a sublayer of either quartz or tungsten, previously deposited on a mica or glass backing. In the P^{32} experiment about eight percent of the recoils escaped in the charged state, whereas in the case of Y^{90} about 87 percent of the recoils escaped as singly charged positive ions. The fact that the recoils from P^{32} have an ionization potential of 10.3 volts as compared with the value of 6.9 volts for the Y^{90} recoils may explain this difference in the relative number of charged recoils.

Several recoil momentum distributions are shown in Fig. (5.32). Since the maximum energy of the recoils should be in the range 41 to 44 ev, it is apparent that the recoils suffered an energy loss of 6 to 9 ev in leaving the surface of the source. The uncertainty in the computed maximum energy of the recoils is due to the spread in the tabulated values of the end point of the beta-spectrum. Sherwin suggests that the yttrium probably was deposited on the catching surface as YO. This would mean that the Zr^{+} recoil ion would have to break a bond with an oxygen atom in order to escape from the monolayer surface as a free ion. As a result of this energy loss, the distribution probably was distorted with too many recoils appearing at the low-energy end.

With the geometry used for the measurements at 180°, $\cos \theta_{ev} \cong \pm 1$ for most of the recoils. In this case the angular correlation $P(W, \theta_{ev})$ of equation (4.8) is approximately proportional to the recoil energy. The corresponding momentum distribution for either the T or A interaction should be much more sharply peaked at the high momentum end than the experimental distribution. Unfortunately, the distortion resulting from the energy losses at the surface of the source prevents a determination of the form of the interaction responsible for the Y^{90} decay. The observed distributions do conflict with the "no neutrino" curve.

5.12. Interpretation of Angular Correlation Experiments

The ft values or "comparative half-lives" of allowed beta-decays are of considerable help in the interpretation of electron-neutrino angular correlation experiments. In many instances the analysis of the experimental ft values will yield values of the coupling constants and matrix elements which can be compared with those deduced from angular correlation measurements. If we neglect cross terms and also omit the pseudoscalar interaction, the ft value is given by:

$$P(W_0) = \frac{\ln 2}{t} = \frac{m^5 c^4}{4\pi^3 \hbar^7} [C_F^2 |\langle 1 \rangle|^2 + C_{GT}^2 |\langle \sigma \rangle|^2] \times f, \qquad (5.16)$$

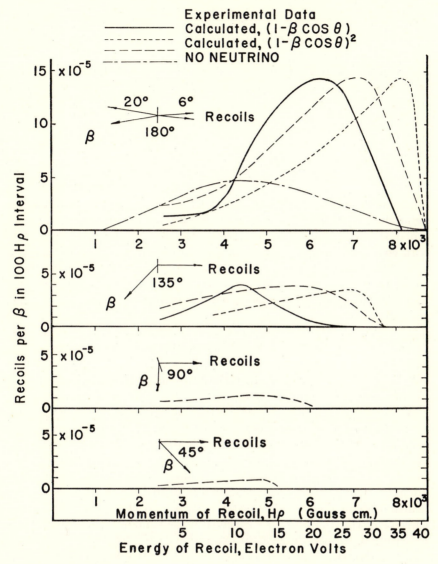

Fig. 5.32. The momentum spectra of the recoil nuclei from the beta-decay of Y^{90}. From Sherwin.[34]

where f is a dimensionless function resulting from the integration of the energy-sensitive part of equation (4.3) over the beta ray spectrum, and t is the partial half-life of the transition. The coupling constants are expressed in terms of the parity conserving and parity non-conserving constants by the relations:

$$C_F^2 = |C_S|^2 + |C_V|^2 + |C_S'|^2 + |C_V'|^2,$$

and   (5.17)

$$C_{GT}^2 = |C_T|^2 + |C_A|^2 + |C_T'|^2 + |C_A'|^2.$$

Winther and Kofoed-Hansen[35] have expressed equation (5.16) in the following form which can be easily plotted:

$$B = ft[(1 - x)|\langle 1 \rangle|^2 + x|\langle \sigma \rangle|^2,$$

with

$$B = \frac{4\pi^3 \hbar^7 \ln 2}{m^5 c^4 (C_F^2 + C_{GT}^2)}, \quad \text{and} \quad x = \frac{C_{GT}^2}{(C_F^2 + C_{GT}^2)}. \qquad (5.18)$$

The experimental ft value combined with calculated matrix elements determined a straight line in a B-x transition. The authors have computed the matrix elements for the transitions between mirror nuclei with closed shells $\pm$ one nucleon, and also for transitions of the type $J = 0$ to $J = 0$, no change of nuclear parity. A least squares method was used to find the common intersection point of the B-x lines determined by these matrix elements and the corresponding ft values. The coordinates of the intersection point were $B_0 = 2787 \pm 70$ and $x_0 = 0.560 \pm 0.012$. The errors quoted are twice the standard error as obtained from the internal consistency of the data. This value of x_0 corresponds to $C_{GT}^2/C_F^2 = 1.28 \pm 0.04$, which is in agreement with the value $1.37 \, {}^{+0.40}_{-0.30}$ computed by Gerhart[40] using the B-x lines for the neutron decay and the O^{14} decay.

A summary of the angular correlation experiments which have yielded quantitative results is presented below.

a. The He⁶ experiments. Since both the Fierz interference terms of equation (4.9) and the $Z/137p$ term of equation (4.10) are negligible in the decay of He⁶, the results of the angular correlation measurements are unambiguous. The most accurately measured value of the electron-neutrino angular correlation factor is $\lambda = +0.34 \pm 0.09$,[15] in close agreement with the value $+\frac{1}{3}$ expected if the Gamow-Teller interaction is tensor. If we use the general expression (4.10) for λ, then we may conclude that the lower limit of the experimental value of λ corresponds to:

$$0 \le \frac{|C_A|^2 + |C_A'|^2}{|C_T|^2 + |C_T'|^2} < 0.2, \qquad (5.19)$$

where the unprimed and the primed constants, respectively, represent the relative strengths of the parity conserving and the parity non-conserving interactions. Unfortunately this result does not show the relative strengths of the parity conserving and non-conserving interactions.

b. The neutron experiment. The interpretation of the electron-neutrino angular correlation in the beta-decay of the neutron is made somewhat ambiguous by the uncertainty in the value of the Fierz interference terms for the Fermi interaction. As mentioned earlier in Section 4.2, Davidson and Peaslee have concluded that the Fierz term for Fermi transitions is $b_F < 0.2$. However, since the experimental value of the angular correlation factor is not strongly affected by the Fierz terms, these probably may be neglected without serious error. Using $|\langle 1 \rangle|^2 = 1$ and $|\langle \sigma \rangle|^2 = 3$ and $C_{GT}^2/C_F^2 = 1.28$, we find that $\lambda_{ST} = +0.06, \lambda_{VA} = -0.06, \lambda_{VT} = +0.47$, and $\lambda_{SA} = -0.47$. The experimental value $\lambda = +0.089 \pm 0.108^8$ is in good agreement with the angular correlation expected for the S, T combination of interactions, but also does not disagree strongly with the value expected for the V, A combination. Instead of comparing the measured value of λ with that computed from the ft value, we may use the experimental value of λ to compute a value of C_{GT}^2/C_F^2. The experimental value of λ together with the calculated matrix elements leads to $C_{GT}^2/C_F^2 = 1.49 \; {}^{+1.44}_{-0.56}$, to be compared with the value 1.28 obtained from the B-x lines. We may conclude that both the angular correlation measurement and the ft values indicate that $C_{GT}^2 > C_F^2$.

c. The Ne^{19} experiments. The experimentally determined values of the electron-neutrino angular correlation are $\lambda = -0.15 \pm 0.2,^{46}$ $\lambda = -0.21 \pm 0.08,^{20}$ and $\lambda = +0.14 \pm 0.13.^{47}$ These values clearly cluster around zero. If we assume that the results of the He^6 experiment prove that the Gamow-Teller part of the beta interaction is tensor, then the range of the angular correlation will be $-1 < \lambda < +\frac{1}{3}$ for the S, T combination or $+\frac{1}{3} < \lambda < 1$ for the V, T combination. It is evident that each of the experimental values of λ lies in the range corresponding to the S, T combination of interactions. Over and above this conclusion it remains relevant to inquire whether the specific value of λ is in agreement with other values of λ expected from the comparative half-life of the transition.

Since the $Ne^{19} \rightarrow F^{19}$ transition is of the mirror type, the matrix element $|\langle 1 \rangle|^2$ is very closely equal to 1.0. When a B-x line constructed with this value of $|\langle 1 \rangle|^2$ and the measured ft for Ne^{19} is drawn through the intersection point at $B_0 = 2787 \pm 70$ and $x_0 = 0.560 \pm 0.012$, the value of the Gamow-Teller matrix element is found to be $|\langle \sigma \rangle|^2 = 1.88 \pm 0.16$. The corresponding predicted values of the angular correlation are:

$$\lambda = -0.06 \pm 0.02 \quad \text{for} \quad S, T, \qquad \lambda = +0.06 \pm 0.02 \quad \text{for} \quad V, A,$$

or

$$\lambda = +0.53 \pm 0.04 \quad \text{for} \quad V, T, \qquad \lambda = -0.53 \pm 0.04 \quad \text{for} \quad A, S.$$

There is excellent agreement between the mean of the experimental values and the correlation predicted for the S, T combination. However, there is also agreement between the experimental mean value and the correlation predicted for the V, A interaction.

For an unambiguous interpretation of the Ne[19] results it is necessary to assume, as in the neutron experiment, that the Fierz interference terms are small enough to be neglected. However, in the case of Ne[19] the part of the angular correlation which shows time reversal dependence may not be negligible in comparison with the rest of the correlation factor. Reference to equation (4.10) will show that the time reversal dependent part of λ is multiplied by $2Ze^2/\hbar cp$. For $Z = 9$ and $p = 2\,mc$, this factor is 9/137. For example, if we apply the two-component neutrino theory and choose $C_A/C_T = 0$, and $C_V/C_S = +1.0 \times i$ (C_V and C_S are $\pi/2$ out of phase since we assume that Fierz terms are zero), then the time reversal part of λ would be -0.15 and the total value of λ would be $+0.09$. It is apparent from this example that the present Ne[19] experiments probably are not sensitive enough to demonstrate the possible existence of a time reversal dependent term unless either the cross product C_V/C_S or $C_A/C_T \cong 1 \times i$. A more extensive analysis of these time reversal terms which may appear in electron-neutrino angular correlation measurements has been carried out by Morita.[29]

d. The A[35] experiments. The A[35] experiment of Herrmannsfeldt et al.[58] yielded the rather unexpected result that the beta interaction appeared to be mainly vector rather than scalar. A comparison of the

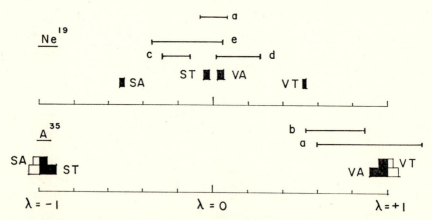

Fig. 5.33. A comparison of the experimental values of the angular correlation coefficients of A[35] and Ne[19], with the values predicted for different pairs of the interaction invariants S, V, A, T. The experimental values indicated by the horizontal bars have been taken from: a) reference 58, b) reference 58, c) reference 20, d) reference 47, e) reference 46. It should be noted that the error given in (e) has been computed on a different basis from those shown in (c) and (d). It is likely that these three experimental values are of about equal weights.

A^{35} and Ne^{19} results is interesting, since both isotopes decay by allowed transitions with the emission of positrons. The experimental values and the computed limits of the angular correlation factors for both transitions are shown in Fig. (5.33). In each case the experimental values are denoted by the solid horizontal bars as mentioned in Section c; the angular correlation for Ne^{19} can be explained by either the S, T or V, A pairs of interactions. If we require the Gamow-Teller part of the interaction to be entirely T in view of the He^6 results, then there is an apparent inconsistency between the experiments on the negatron decay of He^6 and the positron decays of Ne^{19} and A^{35}.

An attempt was made to see whether this inconsistency could be removed by assuming mixtures of the interactions. The best least squares fit of the experimental values of λ with various computed values for the neutron, He^6, Ne^{19}, A^{35} decays occurred when:

$$(|C_S|^2 + |C_S'|^2)/(|C_V|^2 + |C_V'|^2) = 0.43,$$

and

$$(|C_A|^2 + |C_A'|^2)/(|C_T|^2 + |C_T'|^2) = 0.43.$$

Table (5.2) shows the experimental values of λ and also the values computed with these ratios of the coupling constants.

TABLE 5.2

$\lambda(Experiment)$	$\lambda(Computed)$
$n + 0.089 \pm 0.108$	$+0.20$
$He^6 + 0.34 \pm 0.09$	$+0.13$
$Ne^{19} - 0.07 \pm 0.1$	$+0.23$
$A^{35} + 0.70 \pm 0.17$	$+0.37$

It is evident from Table (5.2) that the disagreement between the experimental and corresponding computed values of λ is in most cases more than twice the quoted errors. Apparently it will be difficult to remove this discrepancy with one set of coupling constants.

5.13. Angular Distribution of the Electrons from the Beta-Decay of Polarized Co⁶⁰ Nuclei

Lee and Yang[27] have pointed out that in beta-decay from oriented nuclei, an asymmetry in electron intensity with respect to the nuclear polarization direction would immediately imply non-conservation of parity. Some of the theoretical aspects of this problem have been discussed in Section 4 of Chapter 4. Very fortunately, there is available experimental evidence which provides unequivocal proof that parity is not conserved in beta-decay.

Co^{60} decays by beta emission principally[48] to the second excited state of Ni^{60}. Two prompt gamma rays follow in cascade to the ground state of Ni^{60}. The beta ray spectrum shows an allowed shape with a maximum kinetic energy of 0.309 Mev, and has log $ft = 7.46$. The beta transition is characterized by $J_i = 5 \rightarrow J_f = 4$, with no change in nuclear parity, and accordingly should have only Gamow-Teller nuclear matrix elements. If parity is not conserved in the beta-decay of Co^{60}, the angular distribution of the electrons emitted from polarized nuclei should be given by equations (4.12) and (4.13).

Wu[50] and co-workers have recently found an asymmetry in the angular distribution of the electrons from oriented Co^{60} nuclei, and consequently have demonstrated that parity is not conserved in beta-decay. The nuclei were polarized by the Rose[51]-Gorter[52] method in cerium magnesium (cobalt) nitrate, and the degree of polarization was detected by measuring the anisotropy of the two gamma rays following the beta emission. In this particular experiment two major difficulties had to be overcome. These difficulties originated from the requirements that the beta counter be placed inside the demagnetization cryostat, and that the radioactive nuclei be located in a thin surface layer.

A schematic diagram of the cryostat used in the Co^{60} experiment is shown in Fig. (5.34). To detect electrons, a thin anthracene crystal was placed inside the vacuum chamber about 2 cm above the Co^{60} source. The scintillations were transmitted through a glass window and a lucite light pipe to a photomultiplier at the top of the cryostat. The sensitivity of this counter was found to be independent of temperature and magnetic effects. In order to measure the amount of polarization of the Co^{60}, two NaI gamma ray scintillation counters were installed— one in the equatorial plane and one near the polar position. The observed gamma ray anisotropy was used as a measure of the polarization and the corresponding temperature of the Co^{60} source. The thin radioactive source was prepared by growing a thin crystalline layer containing Co^{60} upon one surface of a single crystal of cerium magnesium nitrate. The thickness of the radioactive layer was about 0.002 inches and contained a few microcuries of activity. The actual measurement of the angular distribution was carried out in the following manner: The temperature of the thin source of Co^{60} was lowered by adiabatic demagnetization of the cerium magnesium nitrate crystal. Immediately after the demagnetization the magnet was opened and a vertical solenoid was raised around the lower part of the cryostat. The beta counting was started and continued during the warm-up period of the source.

The solenoid provided the magnetic field which polarized the Co^{60} nuclei with the nuclear spins pointing either up or down.

The asymmetry in the electron emission as measured by Wu, et al. is shown in Fig. (5.35). The time for the disappearance of the electron asymmetry coincided closely with that of the gamma anistropy, indicating that the beta asymmetry disappeared when the nuclear polarization was destroyed. As mentioned earlier, the large electron asymmetry provides an unequivocal proof that parity is not conserved in beta-decay. The sign of the asymmetry coefficient is negative, and according to equation (4.13) this indicates that the sign for C_T and $C'_T{}^*$ (parity

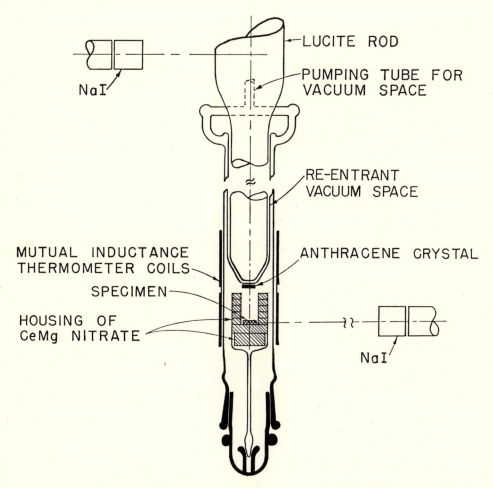

Fig. 5.34. A schematic of the demagnetization cryostat used by Wu et al.[50] in their measurement of the angular distribution of the electrons from the beta-decay of oriented Co^{60} nuclei. The electrons were detected by an anthracene scintillation counter. The Co^{60} nuclei were polarized parallel to the axis of the cylindrical cryostat.

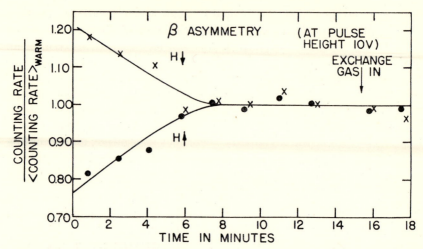

Fig. 5.35. The angular distribution of the electrons emitted in the beta-decay of polarized Co⁶⁰ nuclei. The distribution was measured during the warm-up period of the source. The disappearance of the asymmetry at a time about eight minutes after the start of the measurements coincided with the removal of the polarization of the nuclei. The measured asymmetry indicates that the emission of electrons is favored in the direction opposite to that of the nuclear spin. From the experiment of Wu[50] and co-workers.

conserved and parity non-conserved) must be opposite. This interpretation requires that the term $C_T C_T'^*$ is the dominant factor in the expression for the asymmetry. If we interpret this negative asymmetry in terms of the two-component neutrino theory, then from equations (4.16) and (4.18) we may conclude that the anti-neutrino emitted with the electron is left-handed. The authors estimated that the beta symmetry shown in Fig. (5.35) represented an asymmetry parameter β approximately equal to 0.7. They concluded that this large asymmetry was mainly due to the first term in equation (4.13), and consequently their experiment proved that not only is conservation of parity violated, but invariance under charge conjugation is also violated. This conclusion concerning the invariance of charge conjugation perhaps should be regarded as tentative, since the relative values of the various coupling constants which appear in the asymmetry factor have not been established.†

5.14. Polarization of Electrons in Allowed Beta-Decay

Lee and Yang[53] have suggested that a measurement of the polarization of the electron emitted in beta-decay would be a test of the possible

†Note added in proof.—The interpretation of the Co⁶⁰ experiment is entirely opposite to that given above if we assume that the dominant G.T. interaction is A instead of T. The negative asymmetry coefficient requires that $C_A = +C_A'$, and according to equation (4.17) the anti-neutrino is right-handed.

non-conservation of parity. If parity is not conserved, under certain conditions, the electrons should be longitudinally polarized. Since these experiments should provide information regarding the coupling constants of beta-decay theory, they are similar in this respect to electron-neutrino experiments and will be discussed in this chapter.

If parity is not conserved in beta-decay, then according to Jackson, Treiman, and Wyld,[54] the polarization of a beta-decay electron in a direction n parallel to p is given by:

$$P = \frac{\dfrac{Gp}{W}}{1 + \dfrac{b}{W}}, \qquad (5.20)$$

where b represents the Fierz interference term and has been defined by equation (4.9). The coefficient G, expressed in terms of the parity and non parity-conserving coupling constants introduced by Lee and Yang[27] is given by:

$$G\xi = \pm 2 \text{ Re } [|\langle 1 \rangle|^2 (C_S C_S'^* - C_V C_V'^*) + |\langle \sigma \rangle|^2 (C_T C_T'^* - C_A C_A'^*)]$$

$$+ \frac{2Ze^2}{\hbar cp} Im[|\langle \sigma \rangle|^2 (C_T C_A'^* + C_T' C_A^*) + |\langle 1 \rangle|^2 (C_S C_V'^* + C_S' C_V^*)]; \qquad (5.21)$$

$$\xi = |\langle 1 \rangle|^2 [|C_S|^2 + |C_V|^2 + |C_S'|^2 + |C_V'|^2]$$
$$+ |\langle \sigma \rangle|^2 [|C_T|^2 + |C_A|^2 + |C_T'|^2 + |C_A'|^2], \qquad (5.22)$$

where the upper sign refers to electron emission and the lower to positron emission. The terms within the first square brackets of equation (5.21) vanish if charge conjugation is strictly conserved, and the terms within the second set of square brackets vanish if time reversal is strictly conserved. If $|C_A|$, $|C_A'|$, $|C_V|$, $|C_V'|$ are negligibly small compared to $|C_S|$, $|C_S'|$, $|C_T|$, $|C_T'|$, the detection of possible violations of time reversal invariance by the observation of the momentum dependence of G will be difficult.

The expression for G is greatly simplified when expressed in terms of the two-component neutrino theory of Lee and Yang. In this theory (see equations 4.15, 4.16 and 4.17) the two types of coupling constants are related by $C = -C'$. If we neglect $|C_V|^2$ and $|C_A|^2$ relative to $|C_S|^2$ and $|C_T|^2$ in conformity with the experimental evidence presented in this chapter, then:

$$G = \mp \frac{|\langle \sigma \rangle|^2 |C_T|^2 + |\langle 1 \rangle|^2 |C_S|^2 + \dfrac{2Ze^2}{\hbar cp} Im[|\langle \sigma \rangle|^2 (C_T C_A^*) + |\langle 1 \rangle|^2 (C_S C_V^*)]}{|\langle \sigma \rangle|^2 |C_T|^2 + |\langle 1 \rangle|^2 |C_S|^2}.$$
$$(5.23)$$

If the last term containing the cross products of the coupling constants should prove to be small, then $G = \mp 1$ for allowed beta-decay transitions. According to the work of Sherr and Miller,[55] the Fierz interference term $b \ll 1$, at least for $(\Delta J = \pm 1,$ no) transitions. Under these conditions the longitudinal polarization becomes:

$$P = \mp v_e/c. \tag{5.24}$$

Thus the two-component neutrino theory predicts that electrons from allowed beta-decay transitions should be almost completely polarized.

Polarized electrons can be detected experimentally by atomic scattering experiments, where the spin-orbit coupling terms gives rise to angular asymmetries. Tolhoek[56] shows that the most direct experiment is a measurement of the transverse polarization of an electron by observing the left-right asymmetry in the single scattering from a thin foil of high Z material. This asymmetry is particularly strong at scattering angles between 90° and 150°. The direct observation of the expected longitudinal polarization probably is impossible. However, a longitudinal polarization can be transformed into a transverse type by means of an electrostatic deflector. A deflection of a beam of longitudinally polarized electrons through approximately 90° in an electric field rotates the momentum vector through 90° but not the spin vector. Hence the desired transformation of the polarization is realized. The transversely polarized beam now can be scattered from a thin foil, and the initial longitudinal polarization can be calculated from the measured asymmetry in the scattering.

Frauenfelder[57] and co-workers have recently found that the electrons from the beta-decay of non-oriented Co^{60} nuclei are longitudinally polarized to a degree equal to v/c. A schematic diagram of the apparatus used by Frauenfelder is shown in Fig. (5.36). The electrons from a thin source of Co^{60} were deflected through an angle of about 108° in the electric field between two cylindrical electrodes. The left-right asymmetry of the electrons scattered into the angular interval between 95° and 140° was measured by two end-window Geiger counters. The electrostatic deflector was designed to completely change the polarization of 100 kev electrons from the longitudinal to the transverse type.

For an ideal arrangement, the left-right asymmetry in the scattering would be:

$$L/R = [1 + Pa(\theta)]/[1 - Pa(\theta)],$$

where P is the initial longitudinal polarization of the electrons, and a (θ) is the polarization asymmetry factor for scattering through the angle θ in the analyzer foil. In the actual experiment the determination of P from the L/R ratio involved the following corrections: (1) the asymmetry of the two counters, (2) the finite extension of the scattering foil and counters, (3) incomplete transformation from longitudinal to transverse polarization. The first correction was performed experimentally by using the nearly isotropic scattering from aluminum foils; the second and third corrections were calculated in a first approximation. A correction for depolarization in the source and in the analyzer was neglected.

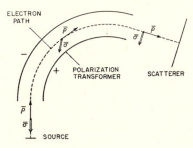

Fig. 5.36a

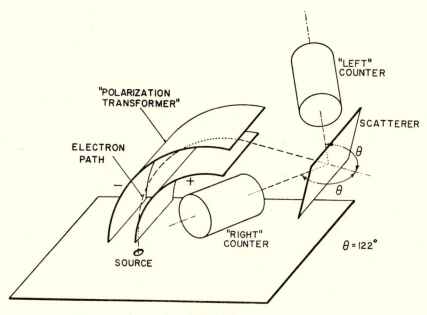

Fig. 5.36b

Fig. 5.36. Schematic diagrams of the apparatus used by Frauenfelder et al.[57] in their measurement of the polarization of the electrons emitted in the beta-decay of Co[60]. The upper diagram shows the general method used to convert the expected longitudinal polarization into the transverse type. Since the electrical field does not change the direction of the spin, the rotation of the momentum vector through approximately 90° changes the polarization from longitudinal to transverse. The lower diagram shows the arrangement of the apparatus used in the measurement of the degree of polarization of the electrons from Co[60].

Preliminary results obtained with the apparatus shown in Figure (5.36) are given in Table (5.3).

TABLE 5.3

The Polarization of Electrons from Co^{60}

Electron Energy		Scattering Foil	Left/Right Asymmetry	Logitudinal Polarization
kev	v/c		L/R	P
50	0.41	0.15 mg/cm² gold	1.03 ± 0.03	−0.04
68	0.47	0.15 mg/cm² gold	1.13 ± 0.02	−0.16
77	0.49	0.05 mg/cm² gold	1.35 ± 0.06	−0.40
77	0.49	0.15 mg/cm² gold	1.30 ± 0.09	−0.35

The following conclusions can be drawn from these results:

(1) The presence of a definite left-right asymmetry proves that parity is violated in this beta-decay process.

(2) The negative sign of the polarization P indicates that each beta particle is polarized in the direction opposite to its momentum.

(3) The values of P are not in disagreement with the value $P = -v_{e/c}$ predicted by the two-component neutrino theory.

If the second and third conclusions are correct, then this experiment proves that a left-handed anti-neutrino is emitted with the electron in beta-decay.

5.15. Notes Added in Proof

During the publication of this book a definite change in the consensus of opinion regarding the form of the beta-decay interaction has occurred as a result of new experimental evidence. This evidence now suggests that the interaction is VA(P) rather than ST(P).

As mentioned earlier in this chapter, the results of the A^{35} experiment indicated that V is the dominant part of the Fermi beta-decay inter-action, at least in the case of positron emission. A confirmation of this result has come from the recent experiment of Goldhaber, Grodzins, and Sunyar[60] on the spirality or helicity of the neutrino. A combined analysis of the circular polarization and resonant scattering of gamma-rays following orbital electron capture measured the helicity of the neutrino. The helicity of the neutrino is left-handed according to this experiment.

This particular method of measuring the helicity of the neutrino may be illustrated by the following example. Consider a nucleus A (spin $J = 0$) which decays by allowed orbital electron capture to an excited state of nucleus B (spin $J = 1$) from which a gamma-ray is emitted to the ground state of $B(J = 0)$. The conditions necessary for resonant scattering are best satisfied for those gamma-rays which are emitted opposite to the neutrino, and have energy comparable to the neutrino and also are emitted before the recoil energy is lost. Since the orbital electrons captured by a nucleus are almost entirely s-electrons, the substates of the daughter nucleus B, formed when a neutrino is emitted in the Z direction, are $m = -1, 0$ if the neutrino has positive helicity, and $m = +1, 0$ if the neutrino has negative helicity. In either case the helicity of the gamma-ray emitted in the $(-Z)$ direction is the same as that of the neutrino. Thus, a measurement of the circular polarization of the gamma rays which are reasonantly scattered by the nucleus B yields directly the helicity of the neutrino.

This experiment was carried out with Eu^{152m} (9.3 hr) which probably has spin 0 and odd parity. This isotope decays to an excited state of Sm^{152} $(1-)$ with emission of neutrinos which have an energy of 840 kev in the most prominent case of K-electron capture. This is followed by an E_1 gamma-ray transition of 960 kev to the ground state $(0+)$. The excited state has a mean life of $(3 \pm 1) \times 10^{-14}$ seconds as determined by Grodzins.[61] Thus, even in a solid source most of the gamma-ray emission takes place before the recoil nucleus has collided with other atoms.

Recent experiments have shown that the magnitude of the longitudinal polarization of the electrons emitted in numerous examples of allowed beta-decay and in several examples of first forbidden decays is, within an experimental accuracy of about ten percent, equal to the maximum amount v_e/c. The sense of the polarization is negative for negative electron emission and positive for positron emission. Reference to equation (5.21) will show that these results require the following choice of coupling constants:

$$C_S = -C'_S \qquad C_T = -C'_T$$
$$C_V = +C'_V \qquad C_A = +C'_A \qquad (5.25)$$

This choice is independent of the relative amounts of the various interactions. An interesting result of this particular choice is that the Fierz interference term of equation (4.9) becomes identically zero.

Since no distinction is made between the neutrino emitted in orbital electron capture and that emitted in positron decay, the Eu^{152} experiment indicates that the neutrino has negative helicity. According to the two-component theory, the anti-neutrino then has a positive helicity. From equation (4.17) we see that the choice of coupling constants is $C_i = +C'_i$. In the case of a Gamow-Teller transition the axial vector (A) interaction is selected by this choice of constants. This is in agreement with the combined results of the Ne^{19} and A^{35} recoil experiments, which show that the dominant beta-decay interaction for positron emitters is VA and not ST.

The electron-neutrino angular correlation in the beta-decay of Ne[23] has been measured recently by Ridley.[62] According to the decay scheme suggested by Penning and Schmidt,[63] the two main transitions go to the ground and first excited states of Na[23]. It appears likely that both transitions are allowed and are pure Gamow-Teller types, and should exhibit identical electron-neutrino angular correlations. The angular correlation factor λ was found to be -0.05 ± 0.10. This value implies a ratio of coupling constants:

$$0.72 \leq \frac{|C_A|^2 + |C_A'|^2}{|C_T|^2 + |C_T'|^2} \leq 2.62.$$

Since the vanishing of the Fierz interference term can no longer be interpreted as an indication that a mixture of axial vector and tensor interactions cannot exist, the Ne[23] results cannot be rejected for this reason. However, there is a strong disagreement between these results and the results of the He[6] experiment of Rustad and Ruby[15] which indicate that the interaction is almost entirely tensor. The latest development in this controversy is a suggestion that the Rustad and Ruby experiment may be in error because of certain experimental difficulties which were not fully appreciated when the original computations were made. In a post deadline paper presented at the New York meeting of the American Physical Society on February 1, 1958 these authors revealed that they had reconsidered this experiment and had discovered two possible sources of error.

Reference to Fig. 5.11 shows that the lower boundary of the source volume used by Rustad and Ruby was defined by a differential pumping aperture. Unfortunately, with this geometry the shape of the electron recoil ion distribution curves at angles near 180° was very sensitive to the extension of the effective source volume in the direction of the recoil ion counter. If this contribution were underestimated then the computed value of the angular correlation factor would have been too negative. A comparison of the computed with the experimental data would yield a value of λ which favored the tensor interaction. Apparently, the true extent of the source volume was not correctly estimated in this experiment. A second source of error resulted from the upward scattering of disintegration electrons from the plastic walls of the differential pumping canal. As in the other type of error this scattering over-emphasized the importance of the source volume inside the pumping canal. Although each of these errors, considered separately, was relatively small, the combination of the two probably caused a relatively large error in the interpretation of the experimental results. Since the difference between the curves corresponding to the tensor or the axial vector interactions is not large, there is a possibility that the interaction in the He[6] decay is A and not T. It is obvious that new He[6] electron-neutrino angular correlation experiments should be carried out to clearly determine the nature of the Gamow-Teller interaction in the case of beta-decay with emission of negative electrons.

Since the source volume used in the Ne^{23} experiment was defined by a differential pumping canal somewhat similar to that used in the He^6 experiment, the possibility exists that the same sources of error were present in both experiments. However, in the evaluation of the angular correlation expected in the Ne^{23} experiment it was assumed that the source strength fell linearly to zero at the far end of the pumping canal. Although this estimate should have accounted for the contributions from most of the source volume, it is entirely possible that a small fraction of the source volume extended beyond the end of the canal. A contribution from this part of the source, if not accounted for in the computations, would have caused the angular correlation coefficient to appear more positive than the correct value.

Burgy et al.[63] have recently measured the asymmetry in the emission of beta particles from the decay of polarized neutrons. This experiment is of great importance to beta-decay theory, since for the first time there is an experimental determination of the relative sign of the Gamow-Teller and Fermi coupling constants. Since the neutron decays with a spin charge $\Delta J = 0$ (no), this is a mixed transition with both Gamow-Teller and Fermi coupling constants. The angular asymmetry factor defined by equations (4.12) and (4.13) becomes, aside from a small Coulomb correction term:

$$\beta\xi = 2\left[\left(\frac{1}{J+1}\right)\mathrm{Re}\,(C_T C_T'^* - C_A C_A'^*)\,|\langle\sigma\rangle|^2 \right.$$

$$\left. + \left(\frac{J}{J+1}\right)^{\frac{1}{2}}\mathrm{Re}\,(C_S C_T'^* + C_S' C_T^* - C_V C_A' - C_V' C_A^*)\,|\langle\sigma\rangle|\,|\langle1\rangle|\right]\frac{p}{W}.$$

$$(5.26)$$

Since the exact values of the nuclear matrix elements are known for the neutron decay, we may compute the values of β to be expected for various combinations of the interaction invariants. We may use the choice of coupling constants shown in relation (5.25) and the relative strengths of the couplings given in Section 5.12. If we assume that the coupling constants are real (time reversal invariance holds), the asymmetry parameter has the limits:

$$\beta = -1.0v_{e/c} \quad \text{for} \quad \text{S} + \text{T} \quad \text{and/or} \quad \text{V} + \text{A},$$

$$\beta = (-0.06 \pm 0.06)v_{e/c} \quad \text{for} \quad \text{S} - \text{T} \quad \text{and/or} \quad \text{V} - \text{A}.$$

$$(5.27)$$

When these two values of the asymmetry parameter are compared with the latest experimental value $\beta = (-0.15 \pm 0.08)v_{e/c}$, it is evident that the beta interaction is S − T and/or V − A. The V − A combination is identical with the pair of interaction forms which are required for an agreement between experimental data and theory in the beta-decay of the μ-meson. Several recent experiments involving the beta-decay of μ-mesons will be reviewed in Section 8.6.

REFERENCES

1. Allen, J. S. *Rev. Sci. Instruments 18*, 739 (1947).
2. Morrish, A. H., and Allen, J. S. *Phys. Rev. 74*, 1260 (1948).
3. Robson, J. M. *Rev. Sci. Instruments 19*, 865 (1948).
4. Chadwick, J., and Goldhaber, M. *Proc. Roy. Soc.* (London), *A151*, 479 (1935).
5. Snell, A. H., and Miller, L. C. *Phys. Rev. 74*, 1217 (1948).
6. Snell, A. H., Pleasonton, F., and McCord, R. V. *Phys. Rev. 78*, 310 (1950).
7. Robson, J. M. *Phys. Rev. 83*, 349 (1951).
8. Robson, J. M. *Phys. Rev. 100*, 933 (1955).
9. Spivac, P. E., Sosnovsky, A. N., Prokofiev, A. Y., and Sokolov, V. S. Geneva Conference 8/P/650, U.S.S.R., 5 July 1955.
10. Wu, C. S., Rustad, B. M., Perez-Mendez, V., and Lidofsky, L. *Phys. Rev. 87*, 1140 (1952).
11. Wigner, E. *Phys. Rev. 51*, 106, 948 (1937); *Phys. Rev. 56*, 519 (1939).
12. Winther, A. *Dan. Mat. Fys. Medd. 27*, No. 2 (1952).
13. Allen, J. S., Paneth, H. R., and Morrish, A. H. *Phys. Rev. 75*, 570 (1949).
14. Allen, J. S., and Jentschke, W. K. *Phys. Rev. 89*, 902 (1953).
15. Rustad, B. M., and Ruby, S. L. *Phys. Rev. 89*, 880 (1953); also *97*, 991 (1955).
16. Christy, R. F., Cohen, E. R., Fowler, W. A., Lauritsen, C. C., and Lauritsen, T. *Phys. Rev. 72*, 698 (1947).
17. Leipunsky, A. I., *Proc. Camb. Phil. Soc. 32*, 301 (1936).
18. Jones, G. A., Phillips, W. R., Johnson, C. M. P., and Wilkinson, D. H., *Phys. Rev. 96*, 547 (1954).
19. Alford, W. P., and Hamilton, D. R. *Phys. Rev. 95*, 1351 (1954).
20. Maxson, D. R., Allen, J. S., and Jentschke, W. K. *Phys. Rev. 97*, 109 (1955).
21. Kofoed-Hansen, O. *Phys. Rev. 96*, 1045 (1954).
22. Snell, A. H., and Pleasonton, F. *Phys. Rev. 100*, 1396 (1955).
23. Snell, A. H. *Bull. Am. Phys. Soc. I*, No. 4, 220 (1956).
24. Enge, H. A. *Phys. Rev. 94*, 730 (1954).
25. Sherwin, C. W. *Phys. Rev. 73*, 1219 (1948); *75*, 1799 (1948); *82*, 52 (1951).
26. Porter, F. T., Wagner, F., and Freedman, M. S. *Phys. Rev. 107*, 135 (1957).
27. Lee, T. D., and Yang, C. N. *Phys. Rev. 104*, 254 (1956).

28. Crane, H. R., and Halpern, J. *Phys. Rev. 53*, 789 (1938); *56*, 232 (1939).
29. Morita, M., and Morita, R. S. *Prog. Theor. Phys. 10*, 345 (1953); *Phys. Rev. 107*, 139 (1957).
30. Thulin, S. *Arkiv. Fysik 4*, 363 (1952).
31. Kofoed-Hansen, O., and Nielson, K. O. *Dan. Mat. Fys. Medd. 26*, No. 7 (1951).
32. Jacobsen, J. C., and Kofoed-Hansen, O. *Dan. Mat. Fys. Medd. 23*, No. 12 (1945); *Phys. Rev. 73*, 675 (1948).
33. Kofoed-Hansen, O., and Kristensen, P. *Dan. Mat. Fys. Medd. 26*, No. 6 (1951); *Phys. Rev. 82*, 96 (1951).
34. Sherwin, C. W. *Phys. Rev. 73*, 1173 (1948).
35. Winther, A., and Kofoed-Hansen, O. *Dan. Mat. Fys. Medd. 27*, No. 14 (1953); *30*, No. 20 (1956).
36. Feenberg, E. Shell Theory of the Nucleus, pp. 119, 140. Princeton University Press, Princeton, 1955.
37. Langer, L. M., and Moffat, R. J. D. *Phys. Rev. 88*, 689 (1952).
38. Hamilton, D. R., Alford, W. R., and Gross, L. *Phys. Rev. 83*, 215 (1951).
39. Schwarzschild, A., Kistner, O., and Rustad, B. M. Columbia University, 1956. Private communication.
40. Gerhart, J. B. *Phys. Rev. 95*, 288 (1954).
41. Wong, C. *Phys. Rev. 95*, 765 (1954).
42. Willard, H. B., and Bair, J. K. *Phys. Rev. 86*, 629 (1952).
43. Green, D., and Richardson, J. R. *Phys. Rev. 101*, 776 (1956).
44. Kistner, O. C., Schwarzschild, A., and Ruby, B. H. *Bull. Am. Phys. Soc. 1*, 30 (1956), paper H-3.
45. Nuclear Data, N. B. S. Circ. 499.
46. Alford, W. P., and Hamilton, A. R. *Phys. Rev. 105*, 673 (1957).
47. Good, M. L., and Lauer, E. J. *Phys. Rev. 105*, 213 (1957).
48. Deutsch, M., and Goldhaber, G. S. *Phys. Rev. 83*, 1059 (1951).
49. Konopinski, E. J. In *Beta and Gamma Spectroscopy*, edited by K. Siegbahn, p. 296. Interscience Publishers, Inc., New York, 1955.
50. Wu, C. S., Ambler, E., Hayward, R. W., Hoppes, D. D., and Hudson, R. P. *Phys. Rev. 105*, 1413 (1957).
51. Rose, M. E. *Phys. Rev. 75*, 213 (1949).
52. Gorter, C. J. *Physica 14*, 504 (1948).
53. Lee, T. D., and Yang, C. N. *Phys. Rev. 105*, 1671 (1957).
54. Jackson, J. D., Treiman, S. B., and Wyld, H. W. *Nuclear Phys. 4*, 206 (1957).
55. Sherr, R., and Miller, R. H. *Phys. Rev. 93*, 1076 (1954).
56. Tolhoek, H. A. *Rev. Mod. Phys. 28*, 277 (1956).
57. Frauenfelder, H., Bobone, R., Von Goeler, E., Levine, N., Lewis,

H. R., Peacock, R. N., Rossi, A., and de Pasquali, G. *Phys. Rev.* *106*, 386 (1957).

58. Herrmannsfeldt, W., Stahelin, P., Maxson, D. W., and Allen, J. S. *Phys. Rev. 107*, 641 (1957).
59. Kistner, O. C., Schwarzschild, A., and Rustad, B. M. *Phys. Rev. 104*, 154 (1956).
60. Goldhaber, M., Grodzins, L., and Sunyar, A. W. *Phys. Rev. 109*, 1015 (1958).
61. Grodzins, L. *Phys. Rev. 109*, 1014 (1958).
62. Ridley, B. W. (to be published).
63. Burgy, M. T., Epstein, R. J., Krohn, V. E., Novey, T. B., Raboy, S., Ringo, G. R., and Telegdi, V. L. *Phys. Rev. 107*, 1731 (1957).

CHAPTER 6

Double Beta-Decay

6.1. Introduction

In general, the present formulation of the theory of beta-decay leads to results which do not depend on the description of the neutrino as either a Dirac or a Majorana particle. However, as discussed in Chapter 2, the shape of an allowed beta-spectrum near the high-energy end point does depend on the nature of the neutrino. This is a small effect and vanishes if the neutrino has zero rest mass. Although experimental studies of the end points of low-energy beta-spectra have set an upper limit for the mass of the neutrino, these measurements have not yielded decisive information regarding the nature of the neutrino.

The phenomenon of double beta-decay provides a means for finding which of the two descriptions of the neutrino is correct. If we assume that the neutrino is a Dirac particle, then the neutrino, ν, is considered to be different from the anti-neutrino, ν^*. The fundamental double beta-decay process can be considered as a two step process:

$$n \rightarrow p + \bar{e} + \nu^*; \qquad (6.1a)$$

followed by

$$n \rightarrow p + \bar{e} + \nu^*. \qquad (6.1b)$$

Since the ν^* emitted in (6.1a) cannot be absorbed in (6.1b), the net result is that two neutrons in the nucleus change to two protons with the emission of two negative electrons and two anti-neutrinos. An important experimental consequence is that the electrons will carry off only part of the available energy of the transition and therefore will have a continuous energy spectrum.

If we assume that there is only one type of neutrino according to the hypothesis of Majorana[1], then the double beta-decay process can take place according to the scheme:

$$n \rightarrow p + \bar{e} + \nu; \qquad (6.2a)$$

and

$$\nu + n \rightarrow p + \bar{e}. \qquad (6.2b)$$

In this case we consider that a virtual neutrino is emitted in the first step and absorbed in the second. The net result is that two neutrons change into two protons with the emission of two electrons. If the energy of the recoiling nucleus is neglected then the two electrons carry off all the energy available for the transition. In contrast to the continuous energy spectrum of the electrons emitted with the Dirac neutrinos, the sum of the kinetic energies of the two electrons should be a sharp line spectrum in the case of the Majorana neutrinos.

In general, pairs of "stable" even-Z, even-N isobars exist. Due to the higher binding energy of the even-even nuclei as a result of the pairing energy of like nucleons, the direct decay of one of these to the other via the intermediate odd-Z, odd-N isobar is energetically forbidden. However, the higher energy even-even nucleus may decay directly to the lower even-even nucleus by double beta-decay. This situation occurs in most of the isobaric doublets (some probable cases are $A = 48$, 110, 116, and 150), and also in the "stable" isobaric triplets ($A = 96$, 124, 130, and 136).

The double beta-decay process is much slower than the usual single beta-decay. For example, if the kinetic energy release is 4 Mev, the half-life for the decay with Dirac neutrinos should be 10^{18} to 10^{19} years, whereas the half-life corresponding to Majorana neutrinos should be 10^{15} to 10^{18} years.

Fireman[2] apparently obtained positive evidence for double beta-decay from Sn^{124}. Numerous other double beta-decay experiments, mostly with negative results have followed the work of Fireman. A list of double beta-decay experiments is given in Table (6.1). Typical experiments selected from this list will be described in a later section of this chapter.

6.2. Theoretical Expectations Concerning Double Beta-Decay

a. Qualitative estimates of the lifetimes. Just as in the case of the theory of single beta-decay, certain aspects of the theory of double beta-decay may be demonstrated by means of a statistical argument. In Chapter 4 we made use of the fact that the transition probability for single beta-decay depends on the density of final energy states available for the electron and the neutrino (or anti-neutrino). Using similar arguments, we expect that the probability of double beta-decay is proportional to the number of energy states available for the two electrons and the two neutrinos. Fireman[3] has carried out this type of analysis in his thesis and we shall follow his treatment of this subject.

A list of symbols which will be used in the formulas is given below. In general the notation is the same as that used in Chapter 4.

w_i = total electron or neutrino energy in mc^2 units.
p = electron momentum in mc units.
q = neutrino energy in mc^2 units = momentum in mc units.

$\epsilon = \epsilon_i - \epsilon_f$ is the nuclear energy release in mc^2 units. $\epsilon - 2$ is the maximum total kinetic energy release = *atomic* mass difference.

Z = nuclear charge, ($\alpha = 1/137$).

R = nuclear radius.

The probability that double beta-decay occurs with electrons s and t, and neutrinos σ and τ, is given by:

$$P \, dW_s \, dW_t \, dW_\sigma \, dW_\tau \tag{6.3}$$
$$= \text{const } p_s W_s p_t W_t q_\sigma W_\sigma q_\tau W_\tau \, dW_s \, dW_t \, dW_\sigma \, dW_\tau.$$

We shall use P_2 to represent the probability of double beta-decay with the emission of two Dirac anti-neutrinos, and P_0 to represent the probability of decay when Majorana neutrinos are emitted and absorbed in a virtual intermediate state.

The condition for conservation of energy for the decay process with Dirac neutrinos is:

$$\epsilon = \epsilon_i - \epsilon_f = W_s + W_t + W_\sigma + W_\tau. \tag{6.4}$$

The transition probability becomes:

$$P_2 \, dW_s \, dW_t \, dW_\sigma \tag{6.5}$$
$$= \text{const } p_s W_s p_t W_t W_\sigma^2 (\epsilon - W_s - W_t - W_\sigma)^2 \, dW_s \, dW_t \, dW_\sigma.$$

Upon integrating over W_σ in (6.5) between the limits of 0 and $\epsilon - W_s - W_t$ we obtain:

$$P_2 \, dW_s \, dW_t = \text{const } \frac{(\epsilon - W_s - W_t)^5}{30} p_s W_s p_t W_t \, dW_s \, dW_t. \tag{6.6}$$

The lifetime for the transition is obtained by integrating (6.6) over the energy spectra of the two electrons:

$$\frac{1}{t_2} = \int_1^{\epsilon-1} \int_1^{\epsilon-W_s} P_2 \, dW_s \, dW_t. \tag{6.7}$$

If the approximations $p_s = W_s/c$ and $P_t = W_t/c$ which are true for high-energy electrons are used, the integration yields:

$$\frac{1}{t_2} = \frac{\text{const}}{1260} (\epsilon - 2)^7 \left[1 + \frac{(\epsilon - 2)}{2} + \frac{(\epsilon - 2)^2}{9} + \frac{(\epsilon - 2)^3}{90} + \frac{(\epsilon - 2)^4}{1980} \right]; \tag{6.8}$$

or approximately:

$$\frac{1}{t_2} = \text{const } 4 \times 10^{-7} (\epsilon - 2)^{11}. \tag{6.9}$$

The condition for conservation of energy for the decay with Majorana neutrinos is:

$$\epsilon = W_s + W_t. \tag{6.10}$$

TABLE 6.1

Double Beta-Decay Experiments

Transition	Kinetic Energy Release	Computed t_0, t_2	Experimental Details	Experimental Half-Life	References
$_{20}Ca^{48} \rightarrow {}_{22}Ti^{48}$	4.3 ± 0.1 Mev[4]	4×10^{15} yrs.	Scintillation counters 35 ft. below ground level	1.6×10^{17} yrs. Positive result?	McCarthy[5]
		4×10^{18} yrs.	Scintillation counters in salt mine 1070 ft. below ground level; Proportional counter	$>2 \times 10^{18}$ yrs.	Awschalom[6]
$_{40}Zr^{96} \rightarrow {}_{42}Mo^{96}$	3.4 ± 0.3 Mev[8]	7×10^{15} yrs.	Scintillation counters	$>2 \times 10^{14}$ yrs. $2 \rightarrow 6 \times 10^{16}$ yrs. Positive result?	Selig,[22] Kohman[20] McCarthy[9]
		9×10^{18} yrs.	Scintillation counters salt mine 1070 ft. below ground level	$>5 \times 10^{17}$ yrs.	Awschalom[6]
$_{42}Mo^{100} \rightarrow {}_{44}Ru^{100}$	2.3 ± 0.2 Mev[24]	6×10^{15} yrs.	Electron sensitive[19] nuclear emulsions 1860 ft. below ground level	1.5×10^{15} yrs. Positive result?	Fremlin and Walters[19]
		2×10^{20} yrs.	Proportional counters Cloud chamber with random triggering, 13,402 acceptable photographs	$>2 \times 10^{16}$ yrs. 3×10^{17} yrs. Positive result?	Kohman,[20] Selig[22] Winter[7]
$_{48}Cd^{116} \rightarrow {}_{50}Sn^{116}$	2.6 ± 0.1 Mev[23]	3×10^{16} yrs.	Cloud chamber with random triggering, 12,352 acceptable photographs	1×10^{17} yrs. Positive result?	Winter[7]
		6×10^{19} yrs.	2 scintillation counters +24 Geiger counters in anti-coincidence	$>4 \times 10^{16}$ yrs.	Detoeuf and Moch[21]
			Nuclear emulsions	$>10^{17}$ yrs.	Fremlin and Walters[19]

(continued on next page)

TABLE 6.1 (continued)

Transition	Kinetic Energy Release	Computed t_0, t_2	Experimental Details	Experimental Half-Life	References
$_{50}Sn^{124} \to {}_{52}Te^{124}$	1.5 ± 0.4 Mev[10]	1 × 10^17 yrs.	2 Geiger-Müller counters in coincidence	4 → 9 × 10^15 yrs.	Fireman[3,11]
	2.0 ± 0.2 Mev[23]	4 × 10^20 yrs.	Cloud chamber	Positive result?	
				>10^16 yrs.	Lawson[12]
			2 screen wall counters in coincidence	>2 × 10^17 yrs.	Kalkstein and Libby[13]
			Cloud chamber triggered by internal counters	>10^17 yrs.	Fireman and Schwarzer[14]
			2 scintillation counters in coincidence	3 → 6 × 10^16 yrs.	Pearce and Darby[18]
			2 scintillation counters in coincidence	>1.5 × 10^17 yrs.	McCarthy[9]
$_{52}Te^{130} \to {}_{54}Xe^{130}$	3.2 ± 0.1 Mev[23]	7 × 10^15 yrs.	Mass spectroscopic analysis of xenon content of a tellurium ore	1.4 × 10^21 yrs.	Inghram and Reynolds[15]
		9 × 10^18 yrs.		Positive result?	
			Mass spectroscopic analysis	3.3 × 10^21 yrs.	Hayden and Inghram[38]
				Positive result?	
$_{60}Nd^{150} \to {}_{62}Sm^{150}$	3.7 ± 0.1 Mev[23]	2 × 10^15 yrs.	Liq. scintillator viewed by two photomultipliers in coincidence	>2 × 10^18 yrs.	Cowan, Harrison, Langer, and Reines[16]
		2 × 10^18 yrs.	Proportional counter	>2 × 10^15 yrs.	Mulholland and Kohman[33]
$_{92}U^{238} \to {}_{94}Pu^{238}$	1.1 Mev[17]	2 × 10^18 yrs.	Chemical separation of Pu from UO₃. Searched for 5.51 Mev alphas from Pu²³⁸	>6 × 10^18 yrs.	Levine, Ghiorso, and Seaborg[17]
		6 × 10^21 yrs.			

In addition there is an added condition imposed upon the two neutrinos:

$$q_\sigma + q_\tau = 0. \tag{6.11}$$

Hence, the neutrinos enter here only in the intermediate state and can be said to be virtually emitted and reabsorbed in the double beta process. Upon use of (6.11) in (6.3), we obtain:

$$P_0 \, dW_s \, dW_t \, dW_\sigma = \text{const } p_s W_s p_t W_t W_\sigma^4 \, dW_s \, dW_t \, dW_\sigma. \tag{6.12}$$

Since the conservation of the momenta of the virtual neutrinos as expressed by (6.11) does not specify the upper limit for W_σ, a value must be obtained from other considerations. A more complete analysis of the problem indicates that if the neutrino wave length becomes of the order of the nuclear dimensions, the neutrino waves begin to average to zero over nuclear dimensions, and as a result, the constant in (6.12) is no longer a constant, but goes to zero. For example, the maximum value of $W_\sigma \approx 52 \; mc^2$, for $R = 7.5 \times 10^{-13}$ cm. Using this upper limit for W_σ, and the approximations $cp_s = W_s$ and $cp_t = W_t$, the integration of (6.12) with respect to W yields:

$$P_0 \, dW_s \, dW_t = \frac{\text{const } (52)^5}{5} \, W_s^2 (\epsilon - W_s)^2 \, dW_s. \tag{6.13}$$

The lifetime for the double beta-decay without the emission of neutrinos is given by:

$$\frac{1}{t_0} = \int_1^{\epsilon-1} P_0 \, dW_s \tag{6.14}$$

$$\cong \text{const } 20(\epsilon - 1)^5. \tag{6.15}$$

When we compare expressions (6.15) and (6.9):

$$\frac{1}{t_2} \cong \text{const } 4 \times 10^{-7} (\epsilon - 2)^{11}, \tag{6.9}$$

we see that the Dirac theory decay rates increase much more rapidly with energy than do the Majorana rates. Since the transition probability increases rapidly with the energy of the neutrinos, the lifetime for the Majorana type of decay will be considerably shorter than that of the Dirac type because of the presence of the higher energy virtual neutrinos required by the Majorana theory. For example, $t_2/t_0 \cong 4 \times 10^2$ if $\epsilon = 10$ (4.1 Mev kinetic energy release).

b. A quantitative treatment of double beta-decay without the emission of neutrinos. The phenomenon of double beta-decay without the emission of neutrinos has been investigated theoretically by Furry[25] and others.[26-29] The most recent treatment of this subject is by Kono-pinski,[30] who made use of Primakoff's results and obtained explicit relations for the lifetimes. Since the complete theoretical treatment is complicated and lengthy, we shall present only the final results from Konopinski's paper. This discussion should show the uncertain factors which prevent accurate estimates of the lifetimes.

In general the double beta-decay transition probability can be calculated by means of second order perturbation theory in accordance with the general scheme of Furry[25]: initial nucleus → intermediate (virtual) nucleus + (virtual) neutrino + electron → final nucleus + two electrons. Konopinski has assumed that only the scalar and tensor couplings contribute to the transition probability as in allowed, single beta-decay. He arrived at the following expression for the decay constant:

$$\lambda_x = \frac{G_x^4}{64\pi^3}\left(\frac{\alpha Z}{R}\right)^2 |\langle f| \ M_x \ i\rangle|^2 \Phi_x(\epsilon - 2). \tag{6.16}$$

The half-life is given by:

$$t_0 = \frac{2.86 \times 10^{-29}}{\lambda} \text{ years}, \tag{6.17}$$

where $\lambda = \lambda_s + \lambda_t$. Values of the scalar and tensor coupling constants G_s and G_t may be derived from single beta-decay experiments. Current values of these constants are:

$$G_s = 2.5 \times 10^{-12}, \tag{6.18}$$
$$G_t^2/G_s^2 = 1.33.$$

The functions Φ_x are integrals over the spectra, and to a rough approximation for $(\epsilon - 2) \approx 4$ to 8 are:

$$\Phi_s = \frac{(\epsilon - 2)^6}{15}; \qquad \Phi_t = \frac{(\epsilon - 2)^6}{5}. \tag{6.19}$$

The nuclear matrix elements $\langle f\ |M_x|\ i\rangle$ probably are the least certain of the various terms of equation (6.16). According to Primakoff, these matrix elements connect the initial and final states of the two isobars. In order to use a single matrix element for this double transition, he made use of "closure" over the intermediate states. This is, in effect, the substitution:

$$\sum_m \langle f\ |M_x|\ m\rangle \cdot \langle m\ |M_x|\ i\rangle = \langle f\ |M_x|\ i\rangle, \tag{6.20}$$

the sums being carried out over the intermediate nuclear states m. This presupposes the existence of a complete enough set of intermediate states within an energy range for which the intermediate neutrino energy must lie, within certain limits. Konopinski has estimated that there is a range of intermediate nuclear energies, roughly 5 to 25 Mev, in which to find a sufficiently complete set of states. He concludes that this makes the "closure" procedure highly plausible.

The matrix elements $\langle f \left| M_x \right| i \rangle$ will contain operators of the order of magnitude unity, as in single beta-decay. On this basis we may assume that:

$$ | \langle f \left| M_x \right| i \rangle |^2 \approx |\langle 1 \rangle|^2 \approx |\langle \sigma \rangle|^2 > \frac{1}{60} , \qquad (6.21) $$

where $\langle 1 \rangle$ and $\langle \sigma \rangle$ are the matrix elements for allowed, single beta-decay as described in Chapter 4. However, if we desire a thorough underestimate of the decay rates, we may assume that only one intermediate state be can found for the transitions, and take:

$$ |\langle f \left| M_x \right| i \rangle|^2 > \left(\frac{1}{60} \right)^2 \qquad (6.22) $$

instead of (6.21). This assumes a damping by $1/60$, the most extreme found in allowed, single beta-decay, for each step of the second order, double beta-decay process.

If we assume that both the scalar and tensor couplings contribute to the decay process, and use the value of the matrix elements given in (6.21), then an adequate estimate of the half-life for double beta-decay without neutrinos is:

$$ t_0 = 1.5 \times 10^{15} \left(\frac{8}{\epsilon - 2} \right)^6 \left(\frac{60}{Z} \right)^2 \left(\frac{A}{150} \right)^{2/3} \text{ years.} \qquad (6.23) $$

Values of t_0 have been computed for various transitions in Table (6.1) and are listed in the third column. In each case t_0 is the shorter lifetime of the two which are listed. In addition a plot of log t_0 as a function of $(\epsilon - 2)$, the kinetic energy release, is shown by the lower curve of Fig. (6.1).

c. A quantitative treatment of double beta-decay with the emission of two neutrinos. The probability of double beta-decay with the mission of two neutrinos has been treated by Goeppert-Mayer[31] on the basis of the original Fermi theory. The computations were carried out by means of second order perturbation theory. According to this theory the decay process will appear as the simultaneous occurrence of two transitions, neither of which separately satisfies the law of conservation of energy. However, energy is conserved between the initial and final states of the system. Konopinski[30] has extended the work of Mayer by including the contributions from the tensor in addition to those from the scalar coupling. He also included terms which represent the angular correlations between the two emitted electrons.

Konopinski derived the following expression for the decay constant:

$$\lambda_x = \frac{G^4}{420\pi^5}(\alpha Z)^2 \{\epsilon_m - \epsilon_i + q + W\}^{-2} |\langle f | M_x | i\rangle|^2$$
$$\cdot [F(\epsilon - 2) - C_x(\epsilon - 2)]. \tag{6.24}$$

$F(\epsilon - 2)$ is a function which arises from a double integration over the energy spectra of the two electrons, and the functions $C_x(\epsilon - 2)$ are corrections which are negligible for $\epsilon = 10$. In the neighborhood of $\epsilon = 10$,

$$F \approx 4 \times 10^{-4}\epsilon^{11}. \tag{6.25}$$

The nuclear matrix elements $\langle f | M_x | i\rangle$ should each consist of the product of two matrix elements as in equation (6.20). In order to obtain lower limits for the estimates of the lifetimes, Konopinski assumed that $|\langle f | M_x | i\rangle|^2 \approx 1$.

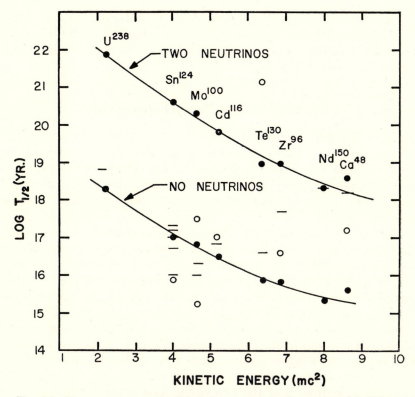

Fig. 6.1. The upper group of solid circles represents the computed half-lives for double beta-decay with the emission of a Dirac neutrino and anti-neutrino. The lower group of solid circles represents the half-lives expected for double beta-decay without the emission of neutrinos. Data from experiments in which a real effect may have been measured are given by the open circles. The short horizontal bars show the lower limits of the half-lives as estimated from statistical considerations in experiments which gave negative results.

The quantity within the curly brackets in equation (6.24) arises from the energy denominators of the second order perturbation expression for the transition probability. The curly brackets signify that some average is to be used. Since the sum of the intermediate electron and neutrino energies has an average about one half the energy release, we may replace $\{W + q\}$ by $\frac{1}{2}\epsilon$. The average value of $\epsilon_m - \epsilon_i$, the difference between the energies of the initial and intermediate nuclear states may be several Mev. For example, if $\epsilon = 10$, $\{\epsilon_m - \epsilon_i + \frac{1}{2}\epsilon\}^{-2} \approx 10^{-2}$.

When the numerical approximations are substituted in equation (6.24), we obtain the following rough formula for the half-life of double beta-decay with the emission of two neutrinos:

$$t_2 = 10^{18}\left(\frac{10}{\epsilon}\right)^{11}\left(\frac{60}{Z}\right)^2 \text{ years.} \tag{6.26}$$

Values of t_2 for various double beta-decay transitions are listed in the third column of Table (6.1). In each case t_2 is the longer lifetime of the two which are listed. A plot of log t_2 as a function of $(\epsilon - 2)$ is shown by the upper curve of Fig. (6.1).

6.3. Double Beta-Decay Experiments

a. experimental procedures. The search for double beta-decay has been carried out by means of various experimental techniques. In general these methods have been based upon the use of either counters, cloud chambers, photographic plates, or chemical separations. Only the more important features of each different experimental technique will be described.

Most of the counter experiments have been designed to demonstrate that double beta-decay occurs according to the Majorana hypothesis. As mentioned earlier, two electrons should be emitted almost exactly coincident in time, and the sum of their energies should be constant. Fireman[3,11] used two counter units, each consisting of two end-window Geiger counters connected to a coincidence circuit. Each pair of counters was arranged so that the two end-windows viewed the opposite surfaces of the sources, which were in the form of thin foils. Surrounding the two sets of end-window counters was a group of fifteen long Geiger counters. These served as guard counters and were connected in anti-coincidence with each of the two end-window units. Both single and coincidence counts were recorded as two source specimens were interchanged between the two counting units. The specimens consisted of equal weights of SnO_2, with one sample containing enriched Sn^{124} and the other, depleted Sn^{124}. Unfortunately, since the amplitude of the pulses from a Geiger counter is nearly independent of the energy of the incident particle, experiments using this type of counter cannot show the shape of the energy spectrum of the electrons.

The experiments of McCarthy[9] and also of Cowan and co-workers[16] are representative of the scintillation counter method. The general features of this type of experiment are the same as those of the Geiger counter experiments. However, there is one important difference between the two methods. Since the amplitude of the pulses from a scintillation counter is energy sensitive, the sum of the energies of two particles detected in coincidence may be displayed by electronically adding the outputs of the two counters. As we have seen, this is impossible when Geiger counters are used. An expanded schematic of the counting and anticoincidence unit used by McCarthy[9] is shown in Fig. (6.2). Transstilbene scintillation crystals were used for both the counting and anti-coincidence units. Simultaneous current pulses from the two scintillation counters were added and recorded by a thirty-channel pulse-height discriminator.

The experiment reported by Cowan and co-workers[16] differed from previous scintillation counter experiments in that large liquid scintilla-

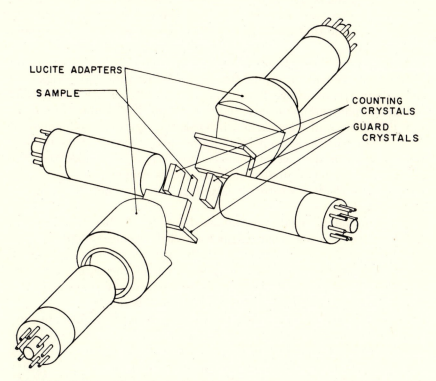

Fig. 6.2. An expanded schematic view of counting and anti-coincidence units used by McCarthy.[9]

tors were used. The source to be investigated was immersed in a liquid scintillator and viewed on each side from a distance of three inches by photomultipliers five inches in diameter. The efficiency of this system for the detection of two electrons, one from each side of the source, was estimated to be thirty percent, for a total kinetic energy release of 4 Mev. A background sample was mounted similarly between a second set of photomultipliers in the same scintillator. Light shields were provided, and an axle permitted periodic interchange of the two samples. Finally, this system of four counters was placed in the center of a 600-liter tank of liquid scintillator viewed by twelve 5-inch photomultipliers, operated in parallel, and providing an effective anti-coincidence shield for the charged particle background.

An example of the use of the cloud chamber technique for the study of double beta-decay is to be found in the experiment of Fireman and Schwarzer.[14] A cylindrical cloud chamber of 12-inch inside diameter and filled with helium and ethyl alcohol at 108 cm of Hg pressure was used. The sample of Sn^{124} in the form of a thin foil was placed across the center of the chamber. A thin-walled Geiger counter was placed on either side of the source, and the chamber was triggered by coincidence pulses from the two counters. The chamber was placed in a uniform magnetic field so that the momenta of acceptable electron tracks could be measured.

Although double electron tracks have been observed in several cloud chamber experiments, it is doubtful as to whether these have originated from double beta-decay. For example, Winter[7] accumulated 12,352 acceptable photographs in his study of the possible double beta-decay of Cd^{116}. There were 24 cases in which negative electron tracks appeared to originate from the same point of source, and three out of the twenty-four events had the correct total energy. Winter concluded that the observed spectrum of double electron events could be caused almost entirely by electron-electron scattering, and by double Compton scattering of photons.

Inghram and Reynolds[15] have looked for the double beta-decay of $Te^{130} \rightarrow Xe^{130}$, using what might be called chemical methods. The rare gases were separated from relatively large amounts of geologically old tellurium ores. The isotopic constitution of the extremely small amount of gas was analyzed in a mass spectrometer. The results of this experiment will be discussed in Section 3.b of this chapter.

Levine, Ghiorso, and Seaborg[17] have used chemical methods in their search for double beta-decay in $U^{238} \rightarrow Pu^{238}$. They separated the Pu fraction from 14 kilograms of six year old UO_3. Since Pu^{238} decays by the emission of 5.51 Mev alphas, and has a half-life of ninety years, this radioactivity could be used as a very sensitive indicator of the possible presence of this isotope in the separated Pu fraction. This method is capable of measuring lifetimes of the order of $10^{18} \rightarrow 10^{19}$ years. A possible weakness of this type of experiment is the fact that the final even-even isobar may have been produced by single rather than double beta-decay. This would require a chain of reactions such as neutron capture followed by one or more single beta-decays leading to the final even-even isobar.

Fremlin and Walters[19] have used electron-sensitive photographic emulsions to look for the presence of double beta-decay in a series of even-even isotopes. A relatively thick sample of each material to be studied was placed against the surface of an emulsion. The actual exposure of the plates was made at a depth of 1860 feet below ground level. The emulsions were scanned for electron tracks after the completion of the exposure. Lifetimes of the order of 10^{17} years probably could be measured by means of this method. However, it is not clear how ordinary single beta-decay could be distinguished from double beta-decay unless the plates were scanned for pairs of electron tracks of common origin.

b. Experimental results. A list of eight possible double beta-decay transitions which have been extensively studied is given in the first column of Table (6.1). The computed half-lives are in the third column and the experimentally determined values are in the fifth column. A graphical display of both the computed and experimentally determined half-lives is shown in Fig. (6.1). In practically all the experiments no real evidence for double beta-decay was discovered. However, in several of the experiments positive results may have been obtained. These "positive" results are indicated by the open circles. In the following sections short summaries of the results from each group of experiments will be given.

$_{20}\text{Ca}^{48} \rightarrow {}_{22}\text{Ti}^{48}$

McCarthy[5] has investigated Ca^{48} by means of the scintillation counter method. The experiment was carried out in the sub-basement of a building 571 feet high. The sub-basement was 35 feet below ground level. The source consisted of approximately 66 milligrams of Ca^{48}. The net number of coincidence counts observed in the region between 2 and 6 Mev is shown in Fig. (6.3). There is definite evidence of a peak at 4.1 ± 0.3 Mev, which is in good agreement with the value of 4.3 ± 0.1 Mev obtained from mass spectroscopic data. A rather poorly defined peak appeared at this same energy in the plot of the total activity against the kinetic energy release. The size of the peak in Fig. (6.3) corresponds to a half-life of $(1.6 \pm 0.7) \times 10^{17}$ years. Reference to Fig. (6.1) shows that this experimental value is about forty times that predicted for decay without neutrinos. Since the theoretical values may be too low by a factor of 1/60, the peak at 4.1 Mev might possibly be due to double beta-decay.

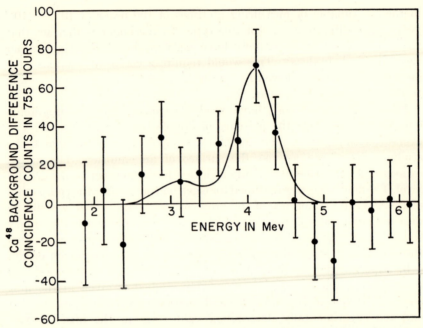

Fig. 6.3. Difference between the coincidence activities of a Ca^{48} and a Ca^{44} sample, for 755 hours of counting for each sample. The curve drawn in indicates the predicted number of coincidences if double beta-decay occurs with a lifetime of 1.6×10^{17} years. The smaller peak at 3.1 Mev was constructed with the assumption of double beta-decay to the first excited state of Ti^{48}. From McCarthy.[5]

Awschalom[6] also has studied Ca^{48}, but with negative results. His equipment was generally similar to that used by McCarthy. There was, however, one significant difference between the two experiments. Awschalom carried out his investigations in a salt mine at a depth of 1070 feet, whereas McCarthy made his measurements at a depth of only 35 feet plus some additional shielding due to a building. In Awschalom's experiment the total coincidence counting rate in the region 3.0 to 4.75 Mev was 4.4 counts per 100 hours. The coincidence rate for the same energy interval in McCarthy's experiment was about 5 counts per 10 hours. A possible explanation for the seemingly positive results of McCarthy may be that additional coincidence counts were recorded when the Ca^{48} sample was in place as a result of the relatively high background. These additional coincidence counts could have been either chance or true coincidences between doubly scattered electrons or between electrons from the double Compton scattering of photons.

$_{40}Zr^{96} \rightarrow {}_{42}Mo^{96}$

McCarthy[9] and also Awschalom[6] have looked for double beta-decay in Zr^{96}. As in the Ca^{48} experiments, McCarthy's measurements were carried out with a relatively small amount of shielding, whereas Awschalom carried out his experiment at a depth of 1070 feet below ground level.

Fig. (6.4) shows the net counting rate observed by McCarthy in the region 2 to 6 Mev. The upper diagram shows the net total activity recorded in 218 hours. The net coincidence activity recorded in a run of 212 hours is shown in the lower diagram. McCarthy computed a half-life of $(6 \pm 2) \times 10^{16}$ years, assuming that the peak in the region of 4 Mev in the plot of total activity was due to double beta-decay. Apparently there is no evidence of a similar peak in the coincidence activity plot.

Awschalom concluded that the results of his experiment showed no evidence for double beta-decay in Zr^{96}. He estimated the lower limit of

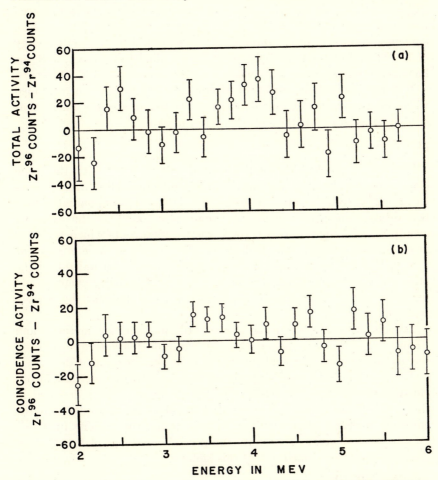

Fig. 6.4. Difference between activities with Zr^{96} and Zr^{94} samples. The difference in the total activity recorded in a run of 218 hours is shown in the upper diagram; the difference in the coincidence activity in 212 hours is shown in the lower diagram. From McCarthy.[9]

the half-life to be 5×10^{17} years. Since this value exceeds the theoretical half-life by a factor of only seventy, the possibility of either type of double beta-decay is not entirely ruled out by this experiment.

$_{42}Mo^{100} \rightarrow {}_{44}Ru^{100}$

Two investigations appear to supply positive evidence for double beta-decay in Mo^{100}. Fremlin and Walters[19] used electron sensitive emulsions at a location 1860 feet below ground level. A week beta activity with a half-life of about 10^{15} years was found in samples containing Mo^{100}. Apparently the existence of pairs of electron tracks was not demonstrated in this experiment. Winter[7] used a cloud chamber with random triggering to investigate Mo^{100}. In 13,402 acceptable photographs there were 35 cases in which two negative electron tracks appeared to originate at the same point on the source. However, there was no evidence for a peak in the number at the expected total kinetic energy of 2.3 Mev. Four out of the 35 events straddled this energy, and the half-life based on these events was estimated to be about 3×10^{17} years.

When the negative results of Kohman[20] and Selig[22] are combined with the two rather doubtful positive results, it appears likely that double beta-decay has not been observed in Mo^{100}.

$_{48}Cd^{116} \rightarrow {}_{50}Sn^{116}$

Winter[7] has looked for double beta-decay in Cd^{116} with the same cloud chamber he used to study Mo^{100}, and obtained similar results. In 12,352 acceptable photographs he found 24 cases of apparent double beta-decay. No concentration of events was found at the expected energy of 2.6 Mev. However, there were three events which occurred at or near this energy. The half-life based upon these three events was 1×10^{17} years. As mentioned in Section 3.a, Winter concluded that the observed energy spectrum of apparent double beta-decay events could have been caused by doubly scattered electrons and also by doubly scattered photons.

Detoeuf and Moch[21] using the scintillation counter method, and Fremlin and Walters[19] using the photographic plate technique were unable to find positive evidence for double beta-decay in Cd^{116}.

The results of the Cd^{116} experiments appear to be somewhat inconclusive. A half-life greater than 10^{17} could not have been detected in these experiments. Since the computed half-lives for decay without the emission of neutrinos probably are lower limits, the half-life for Cd^{116} may be $\approx 10^{18}$ years. Therefore, the possible existence of this type of decay has not been ruled out in this isotope.

$_{50}Sn^{124} \rightarrow {}_{52}Te^{124}$

Fireman[3,11] investigated Sn^{124} using Geiger counters rather than scintillation counters. The general details of his experiment have been described in Section 3.a. He observed a positive effect of 1.94 coincidence counts per hour against a background rate of fifteen counts per hour. Several different runs yielded essentially the same results. He concluded that these results were evidence for double beta-decay with a half-life of $6 \rightarrow 9 \times 10^{15}$ years.

Following Fireman's apparently positive results numerous attempts were made to check his findings. The experiments of Kalkstein and Libby,[13] of Fireman and Schwarzer,[14] and of McCarthy[9] all had negative results, and indicate that the half-life for double beta-decay in Sn^{124} is longer than 10^{17} years. The sensitivity of each of these experiments was sufficiently high to have clearly demonstrated a decay with a lifetime as short as that originally found by Fireman.

It is very difficult to find an explanation for Fireman's apparently positive results. A possible clue is to be found in the statement made by Fireman that the background counting rate depended rather strongly upon the thickness of an absorber placed between the two Geiger counter windows. Although both the actual source and the dummy source had the same weights, a very small difference in the effective scattering cross sections of the two sources could have produced the observed difference in the counting rates. Fireman[14] later concluded that his results were due to impurities in the Sn^{124} source.

$_{52}Te^{130} \rightarrow {}_{54}Xe^{130}$

Inghram and Reynolds[15] used a combination of chemical and mass spectrographic techniques to look for double beta-decay in Te^{130}. Their preliminary experiment was based on a sample of tellurium ore which contained twelve percent by weight of the mineral Bi_2Te_3. Although the age of this sample was known to be $(1.5 \pm 0.5) \times 10^9$ years, there was an uncertainty in the "xenon age" of this material owing to the possibility of comparatively recent crystal alterations by percolating surface waters. No evidence for the presence of Xe^{130} was found in this experiment. The minimum half-life was estimated to be 8×10^{19} years.

A second experiment was carried out with a different sample of tellurium ore. According to geologists it was improbable that the crystals of this Bi_2Te_3 had been affected by a recent alteration. The rare gases were removed and the isotopic constitution analyzed in a mass spectrometer. A definite excess of Xe^{130} over that present in normal xenon was found. Assuming an age of 1.5×10^9 years for the Bi_2Te_3, the excess Xe^{130} corresponded to a double beta-decay with a half-life of 1.4×10^{21} years. A similar experiment by Hayden and Inghram[32] yielded a half-life of 3.3×10^{21} years. These values are well above the computed half-lives for either of the two types of beta-decay. If we assume that the Xe^{130} was formed as a result of double beta-decay, then a relatively large fraction of the xenon must have diffused out of the tellurium ore in some past geological age.

Kohman[20] and Selig[22] have used recent mass spectrographic and radioactivity data to show that the atomic mass difference $Te^{130} - I^{130} = 0.30 \pm 0.11$ Mev. If this mass difference is correct then it is possible that Te^{130} may decay to Xe^{130} by the two successive single beta-decays, $Te^{130} \rightarrow I^{130} \rightarrow Xe^{130}$. This may be the explanation for the apparently positive evidence of double beta-decay in Te^{130}.

$$_{60}Nd^{150} \rightarrow _{62}Sm^{150}$$

A comparison of the computed lifetimes plotted in Fig. (6.1) shows that Nd^{150} probably has the shortest lifetime in the list of the isotopes which have been investigated for double beta-decay. This isotope has been studied by Cowan and his co-workers[16] using the large liquid scintillators described in Section 3.a of this chapter. In recording the data of this experiment pairs of pulses from the pair of photomultiplier tubes viewing the sample were accepted if they were not in coincidence with the guard scintillator. The conditions for acceptance were that each pulse of a pair must lie between 0.35 and 8.0 Mev, and that they must be coincident within 0.5 μsec. The energies of the acceptable pulses were added electronically and displayed on a 100 channel pulse-height analyzer.

Fig. (6.5) shows the energy spectrum recorded in a run of 358 hours using the Nd^{150} source. The residual background was attributed largely to the natural radioactivity of the photomultiplier tubes, and to neutrons arising from interactions of the cosmic radiation with the detector shield. The peak at 5.6 Mev was believed to be due to two successive Compton scatterings of the 7.6 Mev gamma ray from neutron capture in iron. According to the latest atomic mass values compiled by Nier,[23] the total kinetic energy of the two electrons from the double decay of Nd^{150} should be 3.7 ± 0.1 Mev. There is apparently no evidence for a peak outside the statistical fluctuations near 3.7 Mev. The authors estimated the energy resolution of their scintillation spectrometer in the region of interest to be about 0.5 Mev, corresponding to five channels on the pulse-height analyzer. The observed integral counting rate in five channels and associated with one standard deviation from the curve of Fig. (6.5) was 0.048 counts per hour. If the expected peak were within this deviation, the minimum half-life would have been about 2×10^{18} years.

The observed lower limit of the lifetime for double beta-decay in Nd^{150} is about 10^3 times the lifetime predicted for the decay without the emission of neutrinos. The large factor apparently rules out this type of decay and supports the assumption that the neutrino is a Dirac particle.

$_{92}U^{238} \rightarrow {}_{94}Pu^{238}$

Levine, Ghiorso, and Seaborg[17] separated the Pu fraction from 14 kilograms of very pure UO_3; a search was made for the 5.5 Mev alpha particles from the decay of the 90 year Pu^{238}. The results were entirely negative, and the authors estimated the minimum half-life to be 6×10^{18} years. If the total kinetic energy of the two electrons was as low as the quoted value of 1.1 Mev then the half-life for decay without neutrinos would be approximately 2×10^{18} years. This fact would then indicate

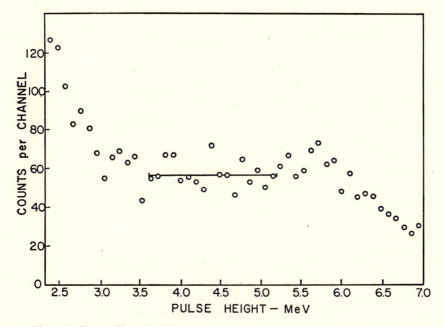

Fig. 6.5. Data collected in 358 hours using a Nd^{150} source. Each point represents the counts recorded at the indicated energy sum. From Cowan, Harrison, Langer, and Reines.[16]

that the sensitivity of this experiment may not have been sufficient for the detection of possible double beta-decay in U^{238}.

6.4. Conclusions

We may conclude from this review of the results of numerous double beta-decay experiments that in most cases the sensitivity of the method has been insufficient for the detection of this type of decay. However, as mentioned in Section 3.b, the measurements of Cowan and co-workers on Nd^{150} and also the work of Awschalom on Ca^{48} and Zr^{96} do appear to have the sensitivity necessary to rule out the type of decay which might occur without the emission of neutrinos. If the sensitivity of the present methods could be increased by a factor of ten, it might be advisable to reinvestigate some of the double beta-decay experiments which have been carried out in the past.

REFERENCES

1. Majorana, E. *Nuovo Cimento 14*, 171 (1937).
2. Fireman, E. L. *Phys. Rev. 74*, 1238 (1948); *75*, 323 (1949).
3. Fireman, E. L. "An Experiment on Double Beta Decay." Unpublished Doctor's Thesis, Princeton University (1948).
4. Collins, T. L., Nier, A. O., and Johnson, W. H. *Phys. Rev. 86*, 408 (1952).
5. McCarthy, J. A. *Phys. Rev. 97*, 1234 (1955).
6. Awschalom, M. *Phys. Rev. 101*, 1041 (1956).
7. Winter, R. G. *Phys. Rev. 99*, 88 (1955).
8. Hogg, B. G., and Duckworth, H. E. *Can. J. Phys. 31*, 942 (1953).
9. McCarthy, J. A. *Phys. Rev. 90*, 853 (1953).
10. Hogg, B. G., and Duckworth, H. E. *Phys. Rev. 86*, 567 (1952).
11. Fireman, E. L. *Phys. Rev. 74*, 1238 (1948); *75*, 323 (1949).
12. Lawson, J. S. *Phys. Rev. 81*, 299 (1951).
13. Kalkstein, M. L., and Libby, W. F. *Phys. Rev. 85*, 368 (1952).
14. Fireman, E. L., and Schwarzer, D. *Phys. Rev. 86*, 451 (1952).
15. Inghram, M. G., and Reynolds, J. H. *Phys. Rev. 76*, 1265 (1949); *78*, 822 (1950).
16. Cowan, C. L., Harrison, F. B., Langer, L. M., and Reines, F. *Nuovo Cimento X*, Vol. 3, 649 (1956).
17. Levine, C., Ghiorso, A., and Seaborg, G. T. *Phys. Rev. 77*, 296 (1950).
18. Pearce, R. M., and Darby, E. K. *Phys. Rev. 86*, 1049 (1952).
19. Fremlin, J. H., and Walters, M. C. *Proc. Phys. Soc.* (London) *A65*, 911 (1952).
20. Kohman, T. P. AEC Report NYO-3626 (1954).
21. Detoeuf, J. F.; and Moch, R. *J. de Phys. et Rad. 16*, No. 12, 987 (1955).
22. Selig, H. Unpublished Doctor's Thesis, Carnegie Institute of Technology (1954); A. E. C. Report NYO-6626.
23. Nier, A. O. Table of Atomic Masses, U. of Minn. (1956).
24. McCarthy, J. A. *Phys. Rev. 95*, 447 (1954).
25. Furry, W. H. *Phys. Rev. 56*, 1184 (1939).
26. Touschek, B. *Z. Physik. 125*, 108 (1948).
27. Tiomono, J. Unpublished Doctor's Thesis, Princeton University (1950).
28. Sliv, L. A. *Zhur. Eksp. i Teort. Fiz. 20*, 11, 1035 (1950).

29. Primakoff, H. *Phys. Rev. 85*, 888 (1952).
30. Konopinski, E. J. U.S.A.E.C. Report LAMS-1949.
31. Goeppert-Mayer, M. *Phys. Rev. 48*, 512 (1935).
32. Hayden, R. J., and Inghram, M. G. Nat. Bur. Stand. Circ. *522*, 189 (1953).
33. Mulholland, G. I., and Kohman, T. P. *Phys. Rev. 85*, 144 (1952).

CHAPTER 7

The Detection of the Free Neutrino

7.1. Introduction

In the past thirty years many attempts have been made to discover some evidence of an interaction between free neutrinos, or anti-neutrinos, and matter. The inverse beta-decay process appears to be the most likely of all the possible interactions. Since this type of decay may be considered as the inverse or reverse of the usual beta-decay process, the cross section of the inverse process can be derived from that of the direct reaction with the aid of the principle of detailed balancing. The reliability of this type of calculation is assured by the dual roles played by the neutrino or anti-neutrino and the electron in the Fermi theory of beta-decay. The cross section calculated for this direct interaction of a neutrino with nuclear matter should be in the neighborhood of 10^{-44} cm^2 per atom.

Since it is apparent that the charge of the neutrino must be zero, there must be some other origin for the possible production of ionization by the neutrino. If the neutrino possesses a magnetic moment, then on passing through matter it will be scattered by atomic electrons. This process would result in the production of ion pairs which could be observed. There is also the possibility that a neutrino may collide with a nucleus with a resulting transfer of linear momentum. With the exception of hydrogen, the kinetic energy of a nucleus recoiling from a collision with a neutrino of several Mev energy would be too small to produce ionization.

7.2. Absorption Experiments

There are a number of experiments which give some rather qualitative information regarding the absorption of neutrinos by matter. These experiments have been reviewed by Crane,[9] but will be summarized here because of their historical value in this field of research.

a. During the early period of the investigation of the phenomena associated with beta-decay, there was considerable speculation as to the origin of the continuous spectra of beta ray energies. Ellis and Wooster[10] in 1927, and Meitner and Orthmann[11] in 1930 used calorimetric methods to measure the heat carried away by the beta rays from RaE. The energy of the beta rays measured by this method and the average energy of the beta rays obtained from the shape of the energy spectrum were equal within the accuracy of the experiments. These results showed that the neutrinos, or whatever the agent was that carried away the missing energy, were not appreciably absorbed in the walls of the calorimeters.

b. An experiment carried out by Wu[12] in 1941 on the internal and external bremsstrahlung of P^{32} shows that neutrinos do not produce ionization effects with an absorption coefficient which might be confused with those of gamma rays. Using a strong source she measured the ionization in an ionization chamber shielded by just enough material to stop the beta rays from the decay of P^{32}. Since this disintegration does not result in gamma radiation, she assumed that the ionization was caused both by inner bremsstrahlung from the P^{32} decay and by external bremsstrahlung due to the stopping of the electrons in the absorber. Her data showed that only 0.004 quanta per disintegration electron came from the P^{32} source. This value was equal to the calculated intensity of the internal bremsstrahlung to within the experimental limits of error. Since the observed ionization apparently was due to the bremsstrahlung, the neutrinos were not observed, and consequently the mass absorption coefficient for neutrinos does not lie within the limits of, say, 10 to 0.001 $gram^{-1}$ cm^{-2}.

c. Wollan[13] has attempted to measure the ionization produced by the passage of neutrinos through hydrogen. A pair of balanced ionization chambers was used in these measurements. One of the chambers was filled with CH_4 at a pressure of about eleven atmospheres, and the other, with argon at a pressure just sufficient to give an equal response to gamma rays. When these chambers had opposite collecting potentials applied to them and were connected together to an electrometer, the apparatus indicated the difference in the two ionization currents. Since the nuclear scattering of neutrinos should produce some ionization in hydrogen but none in argon, the pair of balanced ionization chambers constituted a neutrino detector.

When the ionization chambers were placed outside the shielding of a nuclear reactor, the ion current in the hydrogen-containing chamber could not be measured. Wollan estimated that a neutrino hydrogen-scattering cross section larger than 2×10^{-30} cm^2 would have resulted in a detectable current.

7.3. The Magnetic Moment of the Neutrino

If the neutrino possesses a magnetic moment then on passing through matter it will be scattered by atomic electrons. Ionization of the atoms will occur for sufficiently close encounters if the kinetic energy of the neutrino is greater than the ionization potential of the atoms. Bethe[14] has derived the following expression for the cross section per electron for the production of ion pairs as a result of this process:

$$\sigma = \pi \left(\frac{e^2}{mc^2}\right)^2 n^2 \ln \left(\frac{mc^2}{V_i}\right) \text{ cm}^2, \tag{7.1}$$

where e^2/mc^2 is the classical electron radius, n is the magnetic moment of the neutrino in Bohr magnetons, and V_i is an average ionization potential of the scattering atom. The cross section for air at N.T.P. is approximately:

$$\sigma = 3 \times 10^{-24} n^2 \text{ cm}^2, \tag{7.2}$$

which corresponds to $103n^2$ ion pairs per kilometer path.

Several neutrino absorption experiments give some rather qualitative information regarding the upper limit of the magnetic moment of the neutrino. Chadwick and Lea[20] used a pressure ionization chamber containing N_2 gas to measure the residual ionization from a 100 millicurie $Ra(D + E + F)$ beta ray source surrounded by lead. As the thickness of the lead was increased to 5.8 cm the ionization decreased at a rate which could be interpreted as being due to the absorption of the soft gamma radiation from RaF. They concluded that neutrinos do not produce more than one ion pair per 150 kilometers of path in air at N.T.P. Nahmias[15] carried out the same kind of experiment, but used Geiger counters and five grams of $Ra(D + E + F)$. He observed a counting rate of less than one count per minute when the thickness of the lead was 91 cm. He interpreted this result as an indication that the neutrino does not produce more than one ion pair per 3×10^5 kilometer path in air at N.T.P. From Bethe's formula, $n < 2 \times 10^{-4}$ Bohr magnetons, corresponding to a cross section $\sigma < 10^{-31}$ cm^2 per electron. Barrett,[16] using a Geiger counter filled with helium and an H^3 source of beta rays obtained an upper limit of 4×10^{-34} cm^2 per electron, corresponding to $n < 10^{-5}$ Bohr magnetons.

Hontermans and Thirring[17] used the data of Kulp and Tyron[21] for the residual counting rate in a heavily shielded screen-walled counter. Assuming that the sun provides a flux at the earth of 6×10^{10} neutrinos cm^{-2} sec^{-1}, they computed a cross section less than 10^{-34} cm^2 per electron, giving $n < 10^{-5}$ Bohr magnetons.

Cowan, Reines, and Harrison[18] placed a large liquid scintillation detector near a nuclear reactor. The net counting rate of pulses due to the reactor and in the $0.44 - 2$ Mev energy range was about thirty counts per second. In order to obtain an upper limit for the scattering cross section they attributed this difference to the neutrino. With this assumption the counting rate represented a cross section of about 6×10^{-40} cm^2 per electron. Since the lower limit of the electrons counted

Fig. 7.1. b. Top view of the large liquid scintillation detector used in the free neutrino experiment of Reines and Cowan.[1]

Fig. 7.1. a. Side view of the large liquid scintillation detector used in the free neutrino experiment of Reines and Cowan.[1]

by the scintillation detector was 0.44 Mev, the relation (7.1) for the cross section becomes:

$$\sigma = 7.5 \times 10^{-26} n^2 \text{ cm}^2. \tag{7.3}$$

The cross section observed in this experiment then corresponds to a magnetic moment $n < 10^{-7}$ Bohr magnetons. This result is $\sim 10^3$ short of the theoretical estimate made by Hontermans and Thirring[17] based on the virtual dissociation of the neutrino into neutron, anti-proton, and positron.

7.4. Inverse Beta-Decay Experiments

a. The $\nu^* + p \rightarrow e^+ + n$ Experiment

The reaction:

$$\nu^* + p \rightarrow e^+ + n \tag{7.4}$$

may be considered as the inverse of the direct decay process:

$$n \rightarrow p + \bar{e} + \nu^*, \tag{7.5}$$

which was used in Chapter 1 to define the anti-neutrino. The inverse reaction is obtained from the direct by first transposing the $\bar{e}$ to the left side of equation (7.5), and then reversing the direction of the reaction. In deriving an expression for the cross section for the reaction (7.4), the assumption is made that the interaction between anti-neutrinos and nuclear matter is described by the usual beta-decay interaction. It is also necessary to assume that the nuclear matrix elements are independent of the energy release, and are the same for the direct and inverse reactions.

The transition probability for the inverse reaction is given by essentially the same fundamental expression as described in Chapter 4 by equation (4.2):

$$P(W) = \frac{2\pi}{\hbar} |H'_{if}|^2 \rho(W); \tag{7.6}$$

and the cross section will be given by the total probability divided by the velocity of the incoming neutrino:

$$\sigma_{\text{atom}} = \frac{P(W)}{v_\nu} \cong \frac{P(W)}{c}. \tag{7.7}$$

The main difference between the direct and indirect beta-decay processes is in the density of the final energy states. In ordinary beta-decay the available energy is shared by the anti-neutrino, the electron, and the recoiling nucleus. In the process represented by equation (7.4) the energy is shared by the positron and the recoiling neutron. If we neglect the energy of the neutron then the final state will consist of mono-energetic positrons. When the appropriate expression for $\rho(W)$ has been substituted in equation (7.6), the expression for the transition probability-per-unit-time and per-unit-solid-angle becomes:

$$P(W, \theta_{e\nu}) \, d\omega_{e\nu} = \frac{2\pi}{\hbar} \frac{cp_e W_e |H'_{if}|^2}{(2\pi\hbar c)^3} \, d\omega_{e\nu}, \qquad (7.8)$$

where the notation is the same as that used in Chapter 4. Just as in ordinary beta-decay we may use plane wave functions for the light particles and obtain a nuclear matrix element and a factor representing the angular correlation between the anti-neutrino and the positron. Since we do not wish to measure the angular correlation, this is integrated over the full range of the angle between the positron and anti-neutrino. The expression for the cross section then becomes:

$$\sigma_{\text{atom}} = \frac{G^2 \sum |\langle f \, |M| \, i\rangle|^2}{2\pi} \left(\frac{\hbar}{mc}\right)^2 W_e (W_e^2 - 1)^{1/2} \cdot \text{cm}^2. \qquad (7.9)$$

Values of the cross sections computed from equation (7.9) for several values of the energy of the anti-neutrino are tabulated in Table (7.1). A value of $G^2 \Sigma \, |\langle f \, |M| \, i\rangle|^2 = 5 \times 10^{-23}$ as determined from the decay of the neutron was used in these computations. We see that the cross section increases rapidly with the energy of the anti-neutrinos. The

TABLE 7.1

W_ν	W_e	σ_{atom}
4.5 mc^2	2.0 mc^2	4×10^{-44} cm^2
5.5 mc^2	3.0 mc^2	10×10^{-44} cm^2
10.8 mc^2	8.3 mc^2	90×10^{-44} cm^2

threshold of the reaction should occur at an anti-neutrino energy of approximately 1.8 Mev. The average cross section for anti-neutrinos from a nuclear reactor should be about 6×10^{-44} cm^2 per atom. Lee and Yang[22] suggest that the value of the cross section when computed according to the two-component neutrino theory should be twice as great as in the usual theory. They give the following explanation: The flux of anti-neutrinos is an experimental quantity independent of theory. If the anti-neutrinos moving in a given direction have only one spin state instead of the usual two, by a detailed balancing argument they must have twice the cross section for absorption as the usual anti-neutrinos with two possible spin states.

Reines, Cowan, and co-workers[1,2] have carried out two experiments which apparently have the sensitivity necessary for the measurement of the cross section of the inverse reaction (7.4). The ultimate success of their experiments depended largely upon the following two factors: the use of very large liquid scintillation detectors, and the intense flux of anti-neutrinos emerging from a nuclear reactor. The principle of their method was essentially the same in both experiments. The ν^* entered the large liquid scintillators which contained cadmium in addition to hydrogen. A prompt pulse due to the annihilation of the positron signaled the absorption of an anti-neutrino. A delayed pulse which appeared several micro-seconds after the prompt pulse was produced by the gamma rays, due to the capture of the neutron in the cadmium. To identify the observed signals as neutrino-induced, the energies of the two pulses, their time delay spectrum, the dependence of the signal rate on reactor power, and its magnitude as compared with the predicted rate were used.

The first experiment was carried out at the Hanford, Washington reactor in 1953. The detector was in the form of a right circular cylinder with a diameter of 75 cm and a height of 75 cm. The inside of this cylindrical volume was viewed by 90 photomultiplier tubes. The liquid scintillator was a toluene-terphenyl mixture loaded with cadmium propionate. Two views of this type of detector are shown in Fig. (7.1). Measurements were made with this scintillator located adjacent to the reactor within a shield designed to absorb other radiations from the reactor to which the scintillator was sensitive. Under these conditions they observed an increase in the counting rate of the scintillator of 0.41 ± 0.20 delayed counts per minute when the reactor was operating over that observed with the reactor off. This increase in counting rate, if ascribed to the process $p(\nu^*, e^+)n$ corresponded to a cross section of $(12 \pm 6) \times 10^{-44}$ cm^2/atom. An average cross section of 6×10^{-44} cm^2/atom for this reaction had been computed for the spectrum of anti-neutrinos expected from the reactor. In addition to this reactor-sensitive effect, a relatively large background was observed. These background pulses were independent of the reactor and probably were due to high-energy neutrons. At the conclusion of this experiment it was felt that an identification of the free neutrino probably had been made.

In order to improve the signal-to-background ratio, a second detector was made. The new detector consisted of a multiple-layer arrangement of scintillation counters and target tanks. A schematic diagram of the detector is shown in Fig. (7.2). Three scintillation detectors consisting of rectangular steel tanks containing a purified triethylbenzine solution of terphenyl and a wave length shifter were used. The tops and bottoms of these chambers were thin to low-energy gamma radiation. The interiors of the tanks were painted white, and the liquid scintillator was viewed by 110 five-inch photomultiplier tubes connected in parallel in each tank.

The two target tanks consisted of polyethylene boxes containing a water solution of cadmium chloride. This arrangement provided two essentially independent "triad" detectors, the central scintillation detector being common to both triads. The detector assembly was completely enclosed by a paraffin and lead shield, and was located in an underground room of the reactor building of the A.E.C. Savannah

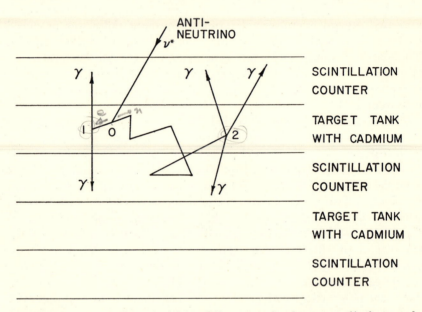

Fig. 7.2. A schematic diagram of the multilayer neutrino detector used in the second experiment of Cowan and Reines.[2] A typical anti-neutrino event is shown. A prompt pulse in both the top and central scintillation counters results from the annihilation of the positron at point 1. A delayed pulse which appears several microseconds later is produced by the gamma rays due to the capture of the neutron at point 2 in the cadmium of the target tank.

River Plant which provided shielding from both the reactor neutrons and gamma rays and also from cosmic rays.

The relatively complete separation of the targets and detectors in the second experiment permitted measurements which were considerably more definite than those made in the first neutrino experiment. For example, the simultaneous presentation of the three tank outputs on the three beam oscilloscopes permitted rejection of penetrating cosmic rays, thus utilizing the two triads as shields for one another.

In order to identify the neutrino-induced reaction (7.4), a detailed investigation was made of each of the factors entering into this relation. Since an understanding of these measurements is necessary for an evaluation of the results of this experiment, they will be summarized here.

1. Dependence on the flux of anti-neutrinos. A reactor power dependent signal was observed which was within five percent of the predicted cross section of 6.3×10^{-44} cm^2 for reaction (7.4). The computed cross section is uncertain by ± 25 percent, largely due to the uncertainty in the shape of the anti-neutrino spectrum. The shape of this spectrum was obtained from recent measurements of the spectrum of beta radiation from fission fragments as measured by C. O. Muehlhause at Brookhaven National Laboratory. The maximum neutrino signal rate was 2.88 ± 0.22 counts-per-hour with a signal-to-background ratio of three to one.

As mentioned earlier in this section, the cross section for the absorption of pile neutrinos should be about 12×10^{-44} cm^2 when computed according to the two-component neutrino theory. Since the experimental value is in excellent agreement with the value computed from conventional neutrino theory, this may indicate that the neutrino should not be described entirely in terms of the two-component theory.

2. Dependence of the signal rate on the number of protons in a target box. This was tested by dilution of the light water solution in a target tank with a heavy water solution, to yield a resultant proton density of half the normal value. Under these conditions the reactor signal fell to half its former value. Although it is likely that there was some anti-neutrino disintegration of the deuterons, this effect probably was very small. Weneser[3] has estimated that the cross section for the disintegration of the deuteron should be approximately 1/30 times that of the proton reaction.

3. The prompt pulses should be due to the radiation from the annihilation of positrons in the target tanks. Analysis of the pulse amplitude spectrum of those gamma rays associated with short time-delay events yielded a spectrum which matched that produced by a positron source placed in a target box. A differential absorption measurement was made by placing lead sheets between a target and a scintillation tank. The introduction of the absorbers caused a reduction of the reactor signal rate as required for events with first pulse gamma rays of 0.5 Mev originating in a target box.

4. The delayed pulse should result from the capture of neutrons in the cadmium target. The time delay spectrum and also the total energy limits of the delayed pulses were in excellent agreement with the values expected for the cadmium concentration used in the target tanks. The removal of the cadmium from the target water resulted in a disappearance of the reactor signal.

5. The effect of fast neutrons from the reactor. The detector shield was surrounded by bags of sawdust saturated with water. This should have reduced by a factor of ten a counting rate due to fast neutrons. No decrease was observed in the reactor signal within the statistical fluctuations quoted in paragraph (1) above.

In view of the detailed investigation of reaction (7.4) and the excellent agreement between the experimental and theoretical values of the cross section, it is apparent that the free neutrino has finally been detected in these two experiments. These experiments do not distinguish between the Dirac and the Majorana description of the neutrino. According to the definitions of the anti-neutrino and neutrino as stated in Chapter 1, a nuclear reactor should be a source of either anti-neutrinos or Majorana neutrinos, and reaction (7.4) should occur in either case. However, as a result of the double beta-decay experiments described in Chapter 6, we may assume that the Dirac picture of the neutrino is correct.

b. The $\nu^* + Cl^{35} \rightarrow S^{35} + e^+$ Experiment

The reaction:

$$\nu^* + Cl^{35} \rightarrow S^{35} + e^+ \tag{7.10}$$

is the inverse of the normal beta-decay:

$$S^{35} \rightarrow Cl^{35} + e^- + \nu^*. \tag{7.11}$$

S^{35} decays with a half-life of 87 days, and emits negative electrons with a maximum kinetic energy of 0.17 Mev. The threshold of reaction (7.10) should occur when the energy of the incident anti-neutrino is about 1.17 Mev. Since $\log ft = 5.2$ for reaction (7.11), and $\log ft = 3.1$ for the beta-decay of the neutron, the cross section for reaction (7.10) should be smaller than the cross section for the $p(\nu^*, e^+)n$ reaction by a factor of approximately 10^{-2}.

An early experimental investigation of the inverse beta-decay process was carried out by Crane[8] in 1939. The source of the anti-neutrinos was one millicurie of mesothorium in equilibrium with its products. The source was placed at the center of a three pound bag of NaCl and left for 90 days. The salt was then dissolved, a sulfur carrier added, and then the sulfur was extracted. No beta activity was found in the sulfur. The sensitivity of the experiment was such that a cross section of 10^{-30} cm^2 per atom could have been detected.

c. The $\nu^* + Cl^{37} \rightarrow A^{37} + e^-$ Experiment

The reaction:

$$\nu + n \rightarrow p + e^- \tag{7.12}$$

may be considered to be the inverse of the direct decay process:

$$e^- + p \rightarrow n + \nu. \tag{7.13}$$

An example of the direct reaction (7.13) is the 34-day electron capture decay of A^{37}. We may express this as:

$$A^{37} + e^- \rightarrow Cl^{37} + \nu, \tag{7.14}$$

with an inverse reaction:

$$Cl^{37} + \nu \rightarrow A^{37} + e^-. \tag{7.15}$$

If the Dirac description of the neutrino is correct, Cl^{37} should not be disintegrated by the anti-neutrinos coming from a nuclear reactor.

The cross section for the inverse reaction (7.15) is given approximately by the relation (7.9) which was derived for the $p(\nu^*, e^+)n$ reaction. For greater accuracy, equation (7.9) should be multiplied by the Coulomb correction factor used in beta-decay theory. The threshold for the $Cl^{37}(\nu, e^-)A^{37}$ reaction should occur when the energy of the incident neutrinos is approximately 0.8 Mev. The cross sections should be about 10^{-2} times those of the $p(\nu^*, e^+)n$ reaction because of the larger ft value for the electron-capture decay of A^{37}. The average cross section of reaction (7.15) for the neutrinos from a nuclear reactor is about 2×10^{-45} cm^2 per atom.

In 1946 Pontecorvo[4] suggested a radiochemical method of studying the $Cl^{37}(\nu, e^-)A^{37}$ reaction. The proposed experiment involved irradiating a large volume of carbon tetrachloride near a nuclear reactor, removing the A^{37} by physical methods, and counting the electron capture decay of this isotope. Alvarez[5] considered in detail the use of this method, and proposed an experiment capable of detecting the expected cross section of 2×10^{-45} cm^2 per atom.

A $Cl^{37}(\nu, e^-)A^{37}$ experiment was carried out by Davis[6] at the Brookhaven reactor. This experiment showed that the major background effects were from the nucleonic component of cosmic radiation and from fast neutrons. In order to reduce the background effects sufficiently to permit the observation of a cross section of 10^{-45} cm^2, it would be necessary to shield the tanks of carbon tetrachloride by an equivalent of twenty feet of earth, and also to keep the flux of neutrons above 1 Mev energy below 10^{-3} neutrons cm^{-2} sec^{-1}.

A second experiment was carried out by Davis[7] at the Savannah River reactor under the required conditions of cosmic ray and fast neutron shielding as determined from his first experiment. The carbon tetrachloride was contained in two 500 gallon tanks which were placed on the lowest floor of the reactor building with the overhead concrete structures providing the additional shielding. The liquid was swept free of argon gas by passing a stream of helium gas through the tanks. The tanks were then closed off under a positive pressure of helium and irradiated with anti-neutrinos from the reactor. At the end of the irradiation the tanks were swept again with helium gas to remove argon. A small measured volume of argon gas, around 0.1 cm³, was introduced into the tanks prior to sweeping, to serve as a carrier and to measure the recovery. The argon gas was separated from the helium in a liquid nitrogen cooled charcoal trap.

After additional purification the argon gas was introduced into a very small Geiger counter. The counting was performed inside an iron shield surrounded by a second shield of mercury. A ring of anti-coincidence counters around the counting chamber reduced the cosmic ray background counting rate. The results of three experiments are summarized in Table (7.2). 500 gallons of CCl_4 were used in the first experi-

TABLE 7.2

Counts per Day

Background	With Sample	Difference	Cross Section for $Cl^{37}(\nu*, e^-)A^{37}$
62.5 ± 2.2	64.9 ± 3.3	2.4 ± 4.0	$(0.7 \pm 1.3) \times 10^{-45}$ cm²
40.9 ± 1.7	46.7 ± 3.0	5.8 ± 3.5	$(0.6 \pm 0.4) \times 10^{-45}$ cm²
32.9 ± 2.6	33.2 ± 2.2	0.3 ± 3.4	$(0.1 \pm 0.6) \times 10^{-45}$ cm²

ment; in all others 1000 gallons were used. An additional experiment has been performed with a counter having a background of thirty counts per day. Results with this counter gave a cross section of $(0.3 \pm 0.4) \times 10^{-45}$ cm². Davis concluded from these experiments that no A^{37} activity was observed above the statistical accuracy of the counting measurements. He tentatively concluded that the cross section for the reaction $Cl^{37}(\nu*, e^-)A^{37}$ is less than 0.9×10^{-45} cm².

During the course of the first experiment performed by Davis, data were obtained which may possibly set an upper limit for the number of neutrinos coming to the earth from the sun. In this experiment a 3900-liter tank of CCl_4 was buried under 19 feet of earth to explore the possibility of producing A^{37} in the tank by processes other than those associated with the nucleonic component of the cosmic radiation. This amount of earth reduced the intensity of the nucleonic component by a factor of about one thousand. The amount of A^{37} activity in the buried tank was found to be less than 0.05 disintegrations per minute, a limit set by the sensitivity of the counter. Two processes are now considered important for energy production in the sun: the proton-proton chain, and the carbon-nitrogen cycle.[19] The neutrinos from the proton-proton chain have a maximum energy which is below the threshold of 0.816 Mev for the $Cl^{37}(\nu, e^-)A^{37}$ reaction, and therefore cannot be detected by this reaction. The neutrinos in the carbon-nitrogen cycle arise from the positron decays of N^{13} and O^{15} which have maximum neutrino energies of 1.24 and 1.68 Mev respectively. Based on this cycle, the flux of neutrinos at the surface of the earth is approximately 6×10^{10} neutrinos cm^{-2} sec^{-1}. The average cross section for these neutrinos is about 3×10^{-46} cm^2 for the $Cl^{37}(\nu, e^-)A^{37}$ reaction. Assuming this cross section, a neutrino flux larger than 1×10^{14} neutrinos cm^{-2} sec^{-1} would have produced a detectable amount of A^{37} in the 3900 liters of CCl_4. We may conclude that this represents an upper limit to the flux of neutrinos in the energy range of 0.8 to 1.7 Mev.

REFERENCES

1. Reines, F., and Cowan, C. L. *Phys. Rev. 92*, 830 (1953).
2. Cowan, C. L., Reines, F., Harrison, F. B., Kruse, H. W., and McGuire, A. D. *Science 124*, No. 3212, 103 (1956).
3. Weneser, J. *Phys. Rev. 105*, 1335 (1957).
4. Pontecorvo, B. *Chalk River Laboratory Report PD-205*, 1948.
5. Alvarez, L. W. *University of California Radiation Laboratory Report UCRL 328* (1949).
6. Davis, R. *Phys. Rev. 97*, 766 (1955).
7. Davis, R., *Bull. Amer. Phys. Soc.* I, No. 4 (April 26, 1956), paper UA5.
8. Crane, H. R. *Phys. Rev. 55*, 501 (1939).
9. Crane, H. R. *Revs. Modern Phys. 20*, 278 (1948).
10. Ellis, C. D., and Wooster, W. A. *Proc. Roy. Soc.* A117, 109 (1927).
11. Meitner, L., and Orthmann, W. *Zeits. f. Physik 60*, 143 (1930).
12. Wu, Chien-Shiung. *Phys. Rev. 59*, 481 (1941).
13. Wollan, E. O. *Phys. Rev. 72*, 445 (1947).
14. Bethe, H. A., *Proc. Camb. Phil. Soc. 31*, 108 (1935).
15. Nahmias, M. E., *Proc. Camb. Phil. Soc. 31*, 99 (1935).
16. Barrett, J. H. *Phys. Rev. 79*, 907 (1950).
17. Hontermans, F. G., and Thirring, W. *Helv. Phys. Acta 27*, 81 (1954).
18. Cowan, C. L., Reines, F., and Harrison, F. B. *Phys. Rev. 96*, 1294 (1954).
19. Salpeter, E. E. *Ann. Rev. Nuc. Sci. 2*, 41 (1953).
20. Chadwick, J., and Lea, D. E. *Camb. Phil. Soc. 30*, 59 (1934).
21. Kulp, J. L., and Tyron, L. E. *Rev. Sci. Inst. 23*, 296 (1952).
22. Lee, T. D., and Yang, C. N. *Phys. Rev. 105*, 1671 (1957).

CHAPTER 8

Meson-Neutrino Reactions

8.1. Introduction

There is rather definite evidence that a very light particle with a spin of $\frac{1}{2}$ is emitted or absorbed in certain processes involving various types of mesons. Although there is no definite proof that this light particle is identical with the neutrino associated with beta-decay, the available evidence does suggest that it has the same general characteristics as the neutrino. The most important processes for our consideration are the following:

$$\pi^{\pm} \rightarrow \mu^{\pm} + \nu, \tag{8.1}$$

$$\mu^{-} + p \rightarrow n + \nu, \tag{8.2}$$

and

$$\mu^{\pm} \rightarrow e^{\pm} + 2\nu. \tag{8.3}$$

The processes (8.1) and (8.2) require the existence of a neutral particle with small rest mass. The decay of the free μ meson into three light particles is closely related to the beta-decay of nuclei. Recent experimental investigations of processes (8.1) and (8.3) have yielded important information regarding the conservation of parity in weak interactions such as those responsible for beta-decay.

8.2. Decay of $\pi^{\pm}$ Mesons

The charged π meson is an unstable particle which decays with a lifetime of 2.6×10^{-8} sec. The principal mode of decay is:

$$\pi^{\pm} \rightarrow \mu^{\pm} + \nu. \tag{8.1}$$

The spin of the $\pi^{\pm}$ meson probably is zero, but a spin of one has not been completely ruled out. In the case of the μ meson a spin of $\frac{1}{2}$ is likely, but a spin of $\frac{3}{2}$ is still a possibility.

In principle, the mass of the neutral particle can be obtained from the balance of momentum and energy in relation (8.1). In order to obtain a solution we need absolute masses of the two mesons which have not been obtained from this same relation. Barkas et al.[1] have derived a value of the $\pi^{\pm}$ masses by measuring the ratio of the momenta and also the ratio of the ranges of mesons and protons within a cyclotron. In this measurement only momentum and range ratios entered into the determination of the meson-proton mass ratios. The meson masses derived from these mass ratios were: π^+ mass $= 277.4 \pm 1.1\ m_e$, and π^- mass $= 276.1 \pm 1.3\ m_e$. The π^+/μ^+ mass ratio has been measured by Birnbaum at al.[2] using a method similar to that employed in the measurement of the π meson-to-proton mass ratio. The results were $\pi^+/\mu^+ = 1.317 \pm 0.004$.

The π^+ meson is known to give rise to a μ^+ meson at the end of its track in a photographic emulsion. Since all available evidence indicates that practically all of the μ^+ mesons have a fixed range of 600 microns in the emulsion, the predominant mode of decay must involve the emission of just one neutral particle in addition to the μ meson. The range of the μ meson is not exactly constant, but shows a distribution resulting from the straggling caused by the statistical nature of the ionization process. The energy corresponding to a range of 600 microns is 4.085 ± 0.044 Mev.

We may assume that a neutral particle of mass m_ν is emitted in reaction (8.1), and attempt to deduce a value for the mass of this particle. From the conservation of energy and momentum we may write:

$$q^2c^2 = T_\mu(T_\mu + 2m_\mu c^2) = (3426 \pm 21)m_e^2 c^4, \qquad (8.4)$$

and

$$W_\nu = m_\pi c^2 - m_\mu c^2 - T_\mu = (59.1 \pm 1.4)m_e c^2. \qquad (8.5)$$

Consequently, the mass of the neutral particle can be found from:

$$m_\nu^2 c^4 = W_\nu^2 - q^2 c^2 = (67 \pm 120)m_e^2 c^4, \qquad (8.6)$$

and

$$m_\nu = (8 \pm 10)m_e. \qquad (8.7)$$

In the equations above, T_μ is the kinetic energy of the μ meson and W_ν is the total energy of the neutral particle. It is obvious that the masses of the mesons are not known with sufficient accuracy to permit more than a rough estimate for m_ν. Instead of invoking a new neutral particle, we may assume that this particle has zero rest mass and is either a neutrino or a photon.

O'Ceallaigh[3] has made an experimental attempt to decide whether the neutral particle involved in the π meson decay is a neutrino or a photon. Some 250 suitable cases of $\pi \rightarrow \mu$ decay in nuclear emulsions were examined for electron pairs which might have been produced by the 30 Mev photons. In a total scan length of 38 cm no such case was observed. A statistical analysis showed that if a photon were emitted, the probability that it would fail to produce a pair under the conditions of the experiment was 4×10^{-3}. This negative result appears to exclude the possibility of appreciable photon emission in the $\pi \rightarrow \mu$ decay process.

8.3. Nuclear Capture of Slow μ⁻ Mesons

Pontecorvo[4] has suggested that μ^- mesons may be captured from the K shell of a mesic atom by a process similar to the capture of orbital electrons. The basic meson capture process is given by:

$$\mu^- + p \rightarrow n + \nu. \tag{8.2}$$

This analogy may imply that there is a direct interaction between the μ, ν fields and the p, n fields of the same form and of the same strength as the interaction required in the theory of beta-decay. The fact that the lifetime computed for the direct meson capture, using the beta-decay coupling constant, is consistent with the experimental value suggests that this hypothesis is correct. However, an alternative solution is that the μ^- meson is absorbed by the proton in a two-stage process. It is assumed that the μ^- may virtually create a π^-, ν pair, and the proton, a π^+, n pair, both of which lead to the same final state of neutron plus a neutrino. In either the direct or indirect coupling case, the nuclear absorption of a slow μ^- meson should be accompanied by the emission of a neutrino.

The experiments of Fry[5] provide evidence that a neutrino is emitted in the nuclear capture of slow μ^- mesons in light elements. Since the energy of the neutrino was in many cases $50 \rightarrow 80$ Mev, it was possible to observe the effect of the neutrino recoil in those reactions in which a neutron was not emitted. He was able to identify reactions of the following type in a photographic emulsion:

$$\mu^- + O^{16} \rightarrow N^{16*} + \nu \rightarrow B^{12} + He^4 + \nu$$
$$\searrow C^{12} + e^- + \nu. \tag{8.8}$$

Reaction (8.8) could be recognized by the B^{12} "hammer tracks," and also by the small departure from collinearity of the B^{12} and alpha particle tracks. An unusual two-prong star caused by a negative μ meson is shown in Fig. (8.1). In this event the energy and momentum balances indicated that the neutrino carried away 70 Mev. Several other μ^- stars were observed, and in each case the neutrino carried away 70 to 80 Mev of energy.

8.4. The μ Meson Beta-Spectrum

Numerous experiments have shown that the μ meson decays according to the scheme:

$$\mu^* \rightarrow e^* + 2\nu. \tag{8.3}$$

Since the energy spectrum of the electrons is continuous, this requires the emission of at least two neutral particles in addition to the electron.

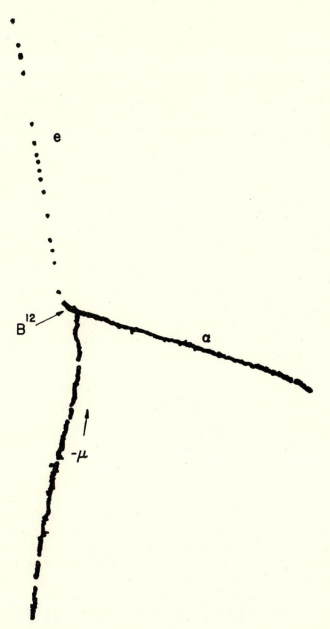

Fig. 8.1. An unusual two-prong star caused by a μ^- meson. The ratio of the lengths of the two nuclear particle tracks, the energy of the beta particle, and the collinearity of the two tracks of the nuclear particles show that the meson was captured by an oxygen nucleus with the emission of a B^{12} fragment, an alpha particle, and a neutrino. From Fry.[5]

If we assume that the two neutral particles each have zero rest mass, then the maximum electron energy will be very nearly $m_\mu c^2/2 = 52.7$ Mev. Recent measurements[6] of the shape of the electron spectrum give an upper-energy cut-off which agrees closely with the value computed from the available energy.

Evidence that the neutral particles resulting from the decay of the μ meson are not photons is provided by the experiments of Hincks and Pontecorvo,[7] and also Sard and Althaus.[8] In these experiments delayed coincidences were sought between the stopped meson and a photon, detected by its materialization in a lead sheet. No evidence for the emission of photons was found in either experiment.

All experimental results suggest that the decay of the μ meson is correctly represented by equation (8.3), and that the $\mu^\pm$ meson is a fermion.

So far there has been no satisfactory explanation of the μ meson decay in terms of an indirect coupling through virtual particles. However, a direct interaction between the four fermions, μ, e, ν, ν analogous to that occurring between p, n, e, ν in beta-decay is suggested by the same interaction strength of about 10^{-49} erg cm^3 in either type of decay. The various types of electron spectra which are predicted by theory for the decay scheme (8.3) were first investigated by Michel,[9] and by Tiomno and Wheeler.[10] The computations were made with an interaction term which was a linear combination of the five interaction forms used in beta-decay theory. In order to carry out the computations it was necessary to assume that the interaction may proceed in three different ways:

(a) $\nu\mu e\nu$,

(b) $\mu\nu e\nu$,

(c) $\mu e\nu\nu$.

Michel was able to obtain a general expression for the shape of the spectrum which applies to each of the three cases and for either distinguishable or indistinguishable neutrinos. If we neglect the rest mass of the electron in comparison with its total energy, the energy distribution is given by:

$$N(W)\,dW = \text{const } W^2\left[3(W_0 - W) + 2\rho\left(\frac{4}{3}W - W_0\right)\right]dW, \qquad (8.9)$$

where W is the total energy of the electron and W_0 is the upper energy cut off. The parameter ρ is a function of the five coupling constants of beta ray theory, and depends on the nature of the neutrino and also on the order in which the fermions interact. When $\rho = 0$, the spectrum goes to zero at the upper limit. If the two neutrinos emitted in the μ

meson decay are distinguishable, $0 \leq \rho \leq 1$. If the two neutrinos are identical then $0 \leq \rho \leq \frac{3}{4}$. Michel has also defined another parameter λ, which is essentially the ratio of the ft values for μ meson decay and the decay of the neutron. λ may be evaluated from the following relation:

$$\lambda = 2^8 B / \mu^5 \tau_\mu \ln 2, \tag{8.10}$$

where the rate constant for beta-decay is:

$$B = \frac{2\pi^3 \hbar^7 \ln 2}{g^2 m^5 c^4} = 2787 \pm 70;$$

μ is the mass of the μ meson, and τ_μ is the mean life for the decay of the μ meson. If we use the current values for μ and τ_μ of $(207 \pm 1)\, m_e$ and $(2.2 \pm 0.02) \times 10^{-6}$ sec, then $\lambda = 1.2 \pm 0.1$. We shall see in the following section that this value of λ may be of some help in selecting the interactions which are responsible for the decay of the μ meson.

The determination of the spectrum of the electrons produced by the decay of the meson has been one of the most extensively investigated problems in particle physics. Some of the difficulties which have been encountered in these experiments have been discussed by Vilain and Williams.[11] One of the principal difficulties was the poor resolution that was inherent in many of the experiments. The earlier data were not accompanied by detailed discussions of the resolution of the respective techniques. An important factor in the overall resolution was the loss of energy suffered by the electrons in traversing the material used to arrest the muons. The experimentally determined values of the parameter ρ which characterizes the shape of the spectrum have varied from 0.23 to 0.72. However, recently reported values of ρ appear to be converging toward a value of about 0.65. Two of the most recently reported experiments will serve as examples of the methods used to measure the shape of the electron spectrum.

Crowe et al.[12] have measured the energy spectrum of the positrons resulting from the decay of μ^+ mesons. The 400 Mev electron beam of the Stanford Mark III linear electron accelerator was allowed to strike a thin lead radiator, and the mixed beam of electrons and x-radiation then passed through a target of lithium. Low-energy mesons produced in the lithium came to rest with relatively uniform density in the target. The positrons which emerged from the target at 90° from the x-ray beam entered a double-focusing magnetic spectrometer. The energy resolution curve of this system resulting from the energy losses of the positrons and the finite widths of the spectrometer openings had a width at half maximum of about 2 Mev. The energy spectrum of the positrons from the lithium target is shown in Fig. (8.2). The authors concluded from these data that the spectrum was characterized by $\rho = 0.50 \pm 0.10$.

Sagane et al.[13] using a spiral orbit, magnetic spectrometer have

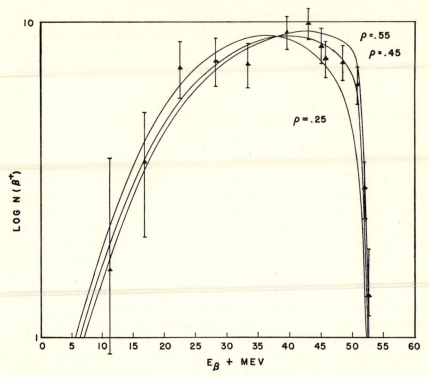

Fig. 8.2. The spectrum of the positrons from the $\mu^+ - e^+ + 2\nu$ decay. The solid curves were obtained by folding the theoretical spectra into the resolution of the apparatus. From Crowe et al.[12]

recently obtained a value of $\rho = 0.62 \pm 0.05$. Apparently their earlier value of 0.23 was due to a systematic error in their experimental procedure.

Sargent et al.[14] have measured the spectrum of the negative electrons from the decay of the μ^- meson. The muons were stopped in the gas of a hydrogen-filled diffusion cloud chamber working at a pressure of 19 atmospheres. The momenta of the decay electrons were determined from their curvature in a magnetic field. The momentum spectrum of the negative electrons is shown in Fig. (8.3). The theoretical curves include corrections required for the resolution of the cloud chamber measurements. In addition, corrections were applied to take care of the fact that approximately 35 percent of the muons decayed from the K shells of mesonic oxygen, nitrogen, or carbon, which were present as impurities in the gas of the cloud chamber. The remainder of the muons decayed from the K shell of mesonic hydrogen. The best fit between the experimental data and the theoretical curves occurred with $\rho = 0.64 \pm 0.10$. The authors have also considered corrections due to certain radiative and non-radiative effects. They have estimated that these effects would increase ρ by approximately 0.04. Their final value of ρ is 0.68 ± 0.11.

The momentum spectrum of the positrons from the decay of the μ^+

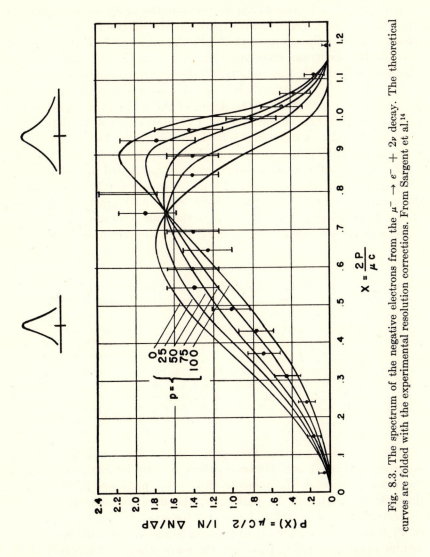

Fig. 8.3. The spectrum of the negative electrons from the $\mu^- \rightarrow e^- + 2\nu$ decay. The theoretical curves are folded with the experimental resolution corrections. From Sargent et al.[14]

meson has been examined by Rosenson,[15] who also used a hydrogen-filled diffusion cloud chamber. The shape parameter found in this experiment was $\rho = 0.68 \pm 0.05$, and is in excellent agreement with the value obtained by Sargent for the spectrum of the electrons from the decay of the μ^- meson.

8.5. The Universal Fermi Interaction

The fact that the rates of muon decay and capture indicate the same interaction strength as does beta-decay and capture has led to the hypothesis of a universal Fermi interaction: the same coupling between any four fermions. The present experimental evidence from beta-decay appears to favor the presence of the S, T, and possibly P interactions. The electron-neutrino angular correlation experiments discussed in Chapter 5 indicate that the relative strength of the S and T interactions is given by $C_s^2/C_t^2 = 0.8$. Michel and Wightman[16] have recently considered the interpretation of the decay of the μ meson in terms of the various combinations of the S, T, and P interactions which have been suggested by other workers in this field. References to the earlier publications on this subject are given in Michel's paper.

According to the experimental evidence presented in section 8.4, we may adopt here $\rho = 0.65 \pm .10$, and $\lambda = 1.2 \pm 0.1$. The values of the parameters ρ and λ computed by Michel and Wightman[16] for each of the three correspondences between the four fermions participating in the μ decay are listed in Table (8.1). From inspection of the contents

TABLE 8.1

Values of the Shape Parameter ρ of μ Decay and of the Ratio λ of the ft Values of μ Decay and β Radioactivity for Various Universal Fermi Interactions

| | $\nu \neq \nu^*$ | | | | | | $\nu \equiv \nu^*$ | | | | | |
| | (a) | | (b) | | (c) | | (a) | | (b) | | (c) | |
	ρ	λ	ρ	λ	ρ	λ	ρ	λ	ρ	λ	ρ	λ
$S + T + P$	0	4/3	3/4	4/3	3/4	4/3	0	4/3	0	1/3	0	1/3
$-S + T - P$	3/4	4/3	0	4/3	3/4	4/3	0	1/3	0	4/3	0	1/3
$\pm S + T \mp P$	3/8	4/3	3/8	4/3	3/4	4/3	0.14	0.92	0.14	0.92	0	1/3

of this table it is evident that processes involving indistinguishable neutrinos are ruled out. However, there is considerable ambiguity in the interpretation of the possible choices for processes involving distinguishable neutrinos. In particular, the decay scheme (c) $\mu e \nu \nu^*$ is characterized by $\rho = 3/4$ and $\lambda = 4/3$ for several combinations of the S, T, P interactions. This is in reasonable agreement with the experimental values of ρ and λ. However, there are other cases in which there is also an agreement between the theoretical and experimental values.

In view of recent theoretical developments, many of the past interpretations of $\pi - \mu$ decay and the decay of the μ meson may be obsolete. Lee and Yang,[17] and also Lee, Oehme, and Yang[18] have suggested that parity, time reversal, and charge conjugation may not be invariant in the "weak" interactions responsible for the beta-decay of nuclei, mesons, and strange particles. These authors have pointed out that if parity is not conserved in the process:

$$\pi \rightarrow \mu + \nu, \tag{8.1}$$

the muon would be polarized in its direction of motion. In the subsequent decay:

$$\mu \rightarrow e + \nu + \nu, \tag{8.3}$$

the angular distribution of the electrons should serve as an analyzer for the polarization of the muon.

Lee and Yang[17] have analyzed a two-component theory of the neutrino which is possible only if parity is not conserved in interactions involving the neutrino. In this theory the neutrino has only one spin state for a given momentum p, the spin being always parallel to p. The spin and momentum of the neutrino together define the sense of a screw. According to this theory the neutrino and anti-neutrino are distinguishable particles, the neutrino corresponding to a right-handed screw, and the anti-neutrino corresponding to a left-handed screw.†

Lee and Yang have applied the two-component neutrino theory to the decay of the muon, which can be expressed as:

$$\mu^- \rightarrow e^- + \nu + \nu^*, \tag{8.11}$$

or

$$\mu^- \rightarrow e^- + 2\nu, \tag{8.12}$$

or

$$\mu^- \rightarrow e^- + 2\nu^*. \tag{8.13}$$

Assuming that the spin of the μ meson is $\frac{1}{2}$, they show that the S, T, P type couplings do not exist for process (8.11). The shape of the electron spectrum for non-polarized μ mesons then would be given by:

$$N(W) \, dW = \text{const } W^2(3W_0 - 2W) \, dW, \tag{8.14}$$

which is characterized by a Michel parameter $\rho = \frac{3}{4}$, which is consistent with the experimental results discussed in Section 8.4. If either process (8.12) or (8.13) prevails, then the electron spectrum is characterized by $\rho = 0$, which is not consistent with experimental evidence. As a result of these arguments we may conclude that (8.11) represents the correct process for the decay of the μ meson.

---

†Note added in proof.—The experimental evidence discussed in section 5.15 now indicates that the helicity of the neutrino is negative and that the helicity of the anti-neutrino probably is positive. The original definitions of the neutrino and anti-neutrino as given by Lee and Yang were arbitrary, and the correct descriptions required experimental verification.

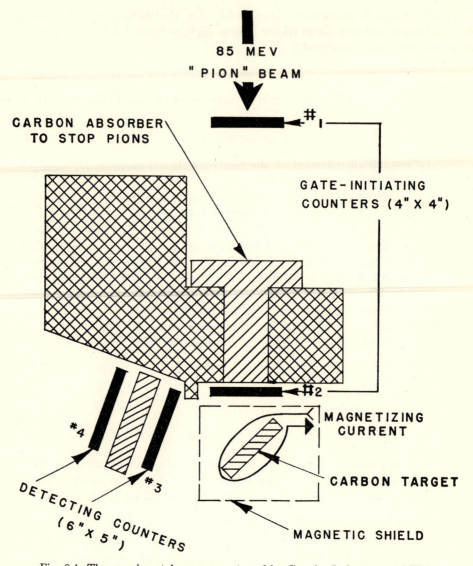

Fig. 8.4. The experimental arrangement used by Garwin, Lederman, and Weinrich[20] to investigate the possible failure of parity conservation in meson decays.

If these new theoretical predictions concerning the decay of the μ meson are substantiated by experimental results, it would appear that nuclear beta-decay is characterized by the S, T, P combination of interactions, whereas the decay of the μ meson involves the V, A combination of interactions. At the present time there is the following experimental evidence that at least some of these predictions have been verified:

The experimental work of Wu and co-workers[19] which has been described in Chapter 5 has established the fact that neither parity nor charge conjugation are invariant in nuclear beta-decay. Garwin and co-workers[20]§ have carried out a beautiful experiment which does indeed show that parity is not conserved in the reactions:

$$\pi^+ \rightarrow \mu^+ + \nu, \tag{8.1}$$

$$\mu^+ \rightarrow e^+ + \nu + \nu^*. \tag{8.3}$$

The experimental arrangement is shown in Fig. (8.4). A beam of 85 Mev π^+ mesons was stopped in a carbon absorber. Some of the μ^+ mesons produced in the carbon absorber according to reaction (8.1) came to rest in a thin carbon target a short distance from the carbon absorber. The stopping of a muon was signalled by a fast coincidence in counters 1 and 2. The subsequent beta-decay of the muon was detected by the electron telescope 3 − 4, which normally recorded electrons with energies greater than 25 Mev.

In this experiment a timing sequence initiated by the pulse from counters 1 − 2 permitted electrons to be recorded in an interval between 0.75 and 2.0 μsecs after the muon had come to rest in the carbon target. It was assumed that the muons created in reaction (8.1) were strongly polarized along their direction of motion. In addition, the assumption was made that the muons were not depolarized during their life in the carbon target. In this case the electrons emitted from the target would have an angular asymmetry about the polarization direction. In the absence of a vertical magnetic field the counter sampled the distribution at $\theta = 100°$. A small magnetic field at the target caused the muons to precess at a rate of $(\mu/sh)H$ radians per second. The probability distribution was carried around with the spin of the precessing muon. In this manner, with a fixed counter system, the entire distribution with angle could be sampled by recording counting rates as a function of the magnetizing current for a given time delay. The results of a typical run are shown in Fig. (8.5). The solid curve is a theoretical fit to a distribution $(1 - \frac{1}{3} \cos \theta)$, with the gyromagnetic ratio for the μ^+ meson taken to be $+2.00$.

The fact that the angular distribution was strongly asymmetric with respect to the spin of the muon showed that many of the muons were polarized along their directions of motion, and that parity and charge conjugation were not conserved in the reactions (8.1) and (8.3). The agreement between the theoretical curve and the experimental data indicated that the gyromagnetic ratio for the μ^+ meson is 2, and strongly

---

§Friedman, J. I. and Telegdi, V. L. *Phys. Rev. 105*, 1681 (1957), also observed the non-conservation of parity in the beta-decay of the μ^+ meson. Their results agreed in sign and magnitude with those of Garwin et al.

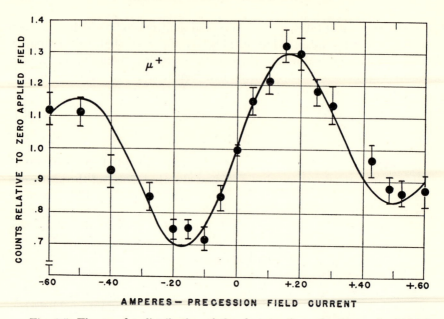

Fig. 8.5. The angular distribution of the electrons from the decay of polarized μ^+ mesons. The solid curve was computed from an assumed distribution $(1 - \frac{1}{3}\cos\theta)$, with counter and gate-width resolution folded in. From Garwin, Lederman, and Weinrich.[20]

suggested that the spin of the μ^+ meson is $\frac{1}{2}$. Later measurements showed that the energy dependence of the angular distribution of the electrons was in rough agreement with that predicted by the two-component neutrino theory of Lee and Yang.[17]

We may conclude that these important theoretical predictions have been partially confirmed and have already made a contribution to our knowledge of the characteristics and behavior of the neutrino. These new developments should provide stimulation for more neutrino experiments.

8.6. Notes Added in Proof

During the publication of this book several new developments have occurred which seem to support the idea of a universal Fermi interaction. Numerous observations of the asymmetry in the emission of electrons in the decay of μ mesons have been made since the first experiments of Garwin and coworkers and of Friedman and Telegdi. These experiments, mainly on μ^+ mesons, have been summarized by Wilkinson.[21] According to the two-component theory of Lee and Yang the angular distribution of the electrons from polarized μ mesons when averaged over the electron energy is given by:

$$\frac{dN}{d\Omega} = \frac{1}{4\pi}\left(1 \pm \frac{\xi}{3}\cos\theta\right),$$ (8.15)

where θ is the angle between the spin direction of the muon and the momentum of the electron. The asymmetry parameter is:

$$\xi = \frac{f_V f_A^* + f_A f_V^*}{|f_V|^2 + |f_A|^2},$$ (8.16)

where the f_i are coupling strengths for the μ beta-decay. The average of a number of experimental results is $|\xi| = 0.87 \pm 0.13$, with the sign of ξ not known since the spin direction of the μ meson was unknown.

The sign of ξ can be determined from an observation of the longitudinal polarization of the electrons from the decay of non-polarized μ mesons. The polarization of the electrons is given by:

$$P = \pm\xi v_e/c \cong \pm\xi,$$ (8.17)

where the $+$ or $-$ sign is to be used for negative or positive μ decay. Culligan et al.[22] have shown that the polarization of the e^+ from μ^+ decay is positive, and consequently, ξ is negative. When the longitudinal polarization results and the asymmetry data are combined in equation (8.16) we obtain $f_A \cong -f_V$ which corresponds to an interaction of the form V − A. This interaction is the same as that deduced from the experiment on the decay of polarized neutrons which was mentioned in Section 5.15. This agreement exists only if we assume that the correct beta-decay interaction is VA and not ST.

Feynman and Gell-Mann,[23] Sudarsham and Marshak,[24] and Sakurai[25] have independently proposed a universal Fermi interaction for all decays which involve four fermions. These theories all predict a VA type of interaction. Feynman and Gell-Mann postulate that in the usual quadrilinear Fermi form of equation (1.3) all fermion fields should be multiplied by the projection operators:

$$a = \tfrac{1}{2}(1 + i\gamma_5) \quad \text{and} \quad \bar{a} = \tfrac{1}{2}(1 - i\gamma_5),$$

where the usual Pauli notation for the Dirac matrices is assumed. The beta-decay interaction then has the form:

$$\sum_i C_i(\overline{a\psi_p}o_i a\psi_n)(\overline{a\psi_e}o_i a\psi_\nu),$$ (8.18)

where $\bar{\psi} = \psi^\dagger\beta$. The properties of the projection operators a and $\bar{a}$ cause the S, T, and P interactions to be removed from equation (8.18) and only A and V survive. Furthermore, both A and V lead to the same coupling constant, and the beta-decay interaction becomes:

$$(8)^{\frac{1}{2}}G(\bar{\psi}_p\gamma_\mu a\psi_n)(\bar{\psi}_e\gamma_\mu a\psi_\nu).$$ (8.19)

The interaction of equation (8.19) is assumed to be universal. For example, the beta-decay of the μ meson can be expressed as:

$$(8)^{\frac{1}{2}}G(\bar{\psi}_e\gamma_\mu a\psi_\mu)(\bar{\psi}_\nu\gamma_\mu a\psi_\nu). \tag{8.20}$$

In the case of beta-decay most of the important features can be successfully explained by the expression (8.19). This expression has the same form as that of the two-component neutrino theory with an interaction $V - A$, which is in agreement with one of the predictions of the polarized neutron experiment described in Section 5.15. The use of the projection operator $a = \frac{1}{2}(1 - i\gamma_5)$ would have made the coupling $V + A$, which is in disagreement with the results of the neutron experiment. The fact that the magnitudes of both the Fermi and the Gamow-Teller coupling constants are required to be equal is in disagreement with the experimentally determined value of $C_{GT}^2/C_F^2 = 1.28 \pm 0.04$ which is given in Section 5.12. A possible explanation for this discrepancy is that the coupling constants are actually equal and that an error has been made in the experimentally determined values. To remove this discrepancy one would have to say that the ft value of 1220 ± 150 sec for the neutron decay was really 1520 sec, and that some uncertain matrix elements in the beta-decay of mirror nuclei were incorrectly estimated.

This theory predicts that the π meson should decay through the radiative process:

$$\pi \rightarrow e + \nu + \gamma \tag{8.21}$$

at a rate which should be less than that of the ordinary $\pi \rightarrow \mu$ decay by a factor of about 1.3×10^{-4}. Experimentally[26] the $\pi \rightarrow e + \nu$ decay has not been observed, indicating that the ratio is less than 10^{-5}. At present, there is no explanation of this discrepancy.

REFERENCES

1. Barkas, W. H., Smith, F. M., and Gardner, E. L. *Phys. Rev. 82*, 102 (1951).
2. Birnbaum, W., Smith, F. M., and Barkas, W. H. *Phys. Rev. 83*, 895 (1951).
3. O'Ceallaigh, C. *Phil. Mag. 41*, 838 (1950).
4. Pontecorvo, B. *Phys. Rev. 72*, 246 (1950).
5. Fry, W. F., *Phys. Rev. 90*, 999 (1953).
6. Crowe, K. M., Helm, R. H., and Tautfest, G. W. *Phys. Rev. 99*, 872 (1955).
7. Hincks, E., and Pontecorvo, B. *Phys. Rev. 73*, 257 (1948).
8. Sard, R., and Althaus, E. *Phys. Rev. 74*, 1364 (1948).
9. Michel, L., *Nature 163*, 959 (1949); *Proc. Phys. Soc. 63A*, 514

(1950); *Phys. Rev. 86*, 814 (1952); Wilson, J. G. *Progress in Cosmic Ray Physics*. New York, Interscience Publishers, 1952; Unpublished Doctor's Thesis, University of Paris, 1953.

10. Tiomno, J., and Wheeler, J. *Rev. Mod. Phys. 21*, 144 (1949).
11. Vilain, J., and Williams, R. *Phys. Rev. 94*, 1011 (1954).
12. Crowe, K. M., Helm, R. H. and Tautfest, G. W. *Phys. Rev. 99*, 872 (1955).
13. Sagani, S., Dudziak, W. F., and Vedder, J. *Bull. Amer. Phys. Soc. 1*, Series II, paper EA14 (April 26, 1956).
14. Sargent, C. P., Rinehart, M., Lederman, L. M., and Rogers, K. C. *Phys. Rev. 99*, 885 (1955).
15. Rosenson, L. *Bull. Amer. Phys. Soc. 2*, Series II, No. 1, paper B-2 (Jan. 30, 1957).
16. Michel, L., and Wightman, A. *Phys. Rev. 93*, 354 (1954).
17. Lee, T. D., and Yang, C. N. *Phys. Rev. 104*, 254 (1956); *Phys. Rev. 105*, 1671 (1957).
18. Lee, T. D., Oehme, R., and Yang, C. N. *Phys. Rev. 106*, 340 (1957).
19. Wu, C. S., Ambler, E., Hayward, R. W. Hoppes, D. D., and Hudson. R. P. *Phys. Rev. 105*, 1413 (1957).
20. Garwin, R. L., Lederman, L. M., and Weinrich, L. M. *Phys. Rev. 105*, 1415 (1957).
21. Wilkinson, D. H. *Il Nuovo Cimento VI*, 517 (1957).
22. Culligan, G., Frank, S. G. F., Holt, J. R., Kluyver, J. C., and Massam, T., *Nature 180*, 751 (1957).
23. Feynman, R. P., and Gell-Mann, M. *Phys. Rev. 109*, 193 (1958).
24. Sudarsham, E. C. G., and Marshak, R. E., (to be published).
25. Sakurai, J. J., (to be published).
26. Lattes, C. and Anderson, H. L. *Il Nuovo Cimento* (to be published).

INDEX

INDEX